Practical Data Engineering with Apache Projects

Solving Everyday Data Challenges with Spark, Iceberg, Kafka, Flink, and More

Dunith Danushka

Practical Data Engineering with Apache Projects: Solving Everyday Data Challenges with Spark, Iceberg, Kafka, Flink, and More

Dunith Danushka
Crewe, Staffordshire, UK

ISBN-13 (pbk): 979-8-8688-2141-7   ISBN-13 (electronic): 979-8-8688-2142-4
https://doi.org/10.1007/979-8-8688-2142-4

Copyright © 2025 by Dunith Danushka

This work is subject to copyright. All rights are reserved by the Publisher, whether the whole or part of the material is concerned, specifically the rights of translation, reprinting, reuse of illustrations, recitation, broadcasting, reproduction on microfilms or in any other physical way, and transmission or information storage and retrieval, electronic adaptation, computer software, or by similar or dissimilar methodology now known or hereafter developed.

Trademarked names, logos, and images may appear in this book. Rather than use a trademark symbol with every occurrence of a trademarked name, logo, or image we use the names, logos, and images only in an editorial fashion and to the benefit of the trademark owner, with no intention of infringement of the trademark.

The use in this publication of trade names, trademarks, service marks, and similar terms, even if they are not identified as such, is not to be taken as an expression of opinion as to whether or not they are subject to proprietary rights.

While the advice and information in this book are believed to be true and accurate at the date of publication, neither the authors nor the editors nor the publisher can accept any legal responsibility for any errors or omissions that may be made. The publisher makes no warranty, express or implied, with respect to the material contained herein.

    Managing Director, Apress Media LLC: Welmoed Spahr
    Acquisitions Editor: Anandadeep Roy
    Coordinating Editor: Jessica Vakili

Distributed to the book trade worldwide by Springer Science+Business Media New York, 1 New York Plaza, New York, NY 10004. Phone 1-800-SPRINGER, fax (201) 348-4505, e-mail orders-ny@springer-sbm.com, or visit www.springeronline.com. Apress Media, LLC is a Delaware LLC and the sole member (owner) is Springer Science + Business Media Finance Inc (SSBM Finance Inc). SSBM Finance Inc is a **Delaware** corporation.

For information on translations, please e-mail booktranslations@springernature.com; for reprint, paperback, or audio rights, please e-mail bookpermissions@springernature.com.

Apress titles may be purchased in bulk for academic, corporate, or promotional use. eBook versions and licenses are also available for most titles. For more information, reference our Print and eBook Bulk Sales web page at http://www.apress.com/bulk-sales.

Any source code or other supplementary material referenced by the author in this book is available to readers on GitHub (https://github.com/Apress). For more detailed information, please visit https://www.apress.com/gp/services/source-code.

If disposing of this product, please recycle the paper

Table of Contents

About the Author xi

About the Technical Reviewer xiii

Acknowledgments xv

Introduction xvii

Part I: Data Lakehouses, Iceberg, Batch ETL, and Orchestration 1

Chapter 1: Foundational Data Engineering Concepts 3

What Is Data Engineering? 3

    Operational Systems vs. Analytical Systems 3

Role of a Data Engineer 5

The Data Engineering Lifecycle 6

    Data Ingestion 7

    Data Processing 7

    Data Storage 8

    Data Serving 8

The Apache Software Foundation and Data Engineering 10

    Apache's Contribution to Big Data Ecosystem 11

Introducing OneShop – Case Study for the Book 12

Roadmap 14

TABLE OF CONTENTS

Prerequisites 16
   Docker Environment 16
   Project Code Repository 17
   Programming Knowledge 18
Summary 19

Chapter 2: Implementing the Data Lakehouse 21

OneShop Data Lakehouse 22
The Problems of Data Warehouses and Data Lakes 23
What Is a Data Lakehouse? 24
What Is a Table Format? 25
Before You Begin 26
Breakdown of the Docker-Compose File 26
Deconstructing the Spark-Iceberg Dockerfile 30
Running Everything 34
Creating a Database and an Iceberg Table 38
Adding Records to the Table 41
Querying the Table with Pyiceberg 42
Working with Pyiceberg CLI 45
Summary 47

Chapter 3: Batch ETL Pipeline with Apache Spark 49

Hydrating the OneShop Lakehouse 50
Medallion Architecture 51
   Bronze Layer (Raw) 51
   Silver Layer (Validated) 51
   Gold Layer (Business) 51

TABLE OF CONTENTS

Before You Begin 52
Docker Setup Overview 52
Running the Setup 56
Modeling Iceberg Tables in the Lakehouse 58
    Creating Namespaces 60
    Modeling Bronze Tables 60
    Modeling Silver Tables 64
Creating Iceberg Tables by Running the Notebook 68
    Validating Table Creation 69
Loading Iceberg Tables from Postgres and MinIO 71
    Loading Postgres Tables 71
    Loading PageView Events from MinIO 74
Data Cleaning, Denormalization, and Enrichment with PySpark 77
Verify Silver Tables 84
    Option 1: Using a Spark Notebook 84
    Option 2: Using Pylceberg CLI 84
    Confirming Successful Data Transformation 85
Summary 85

Chapter 4: Data Visualization with Apache Superset 87

KPIs for OneShop 87
Before You Begin 88
Docker Setup Overview 88
Running Everything 91
Create Gold Tables with Trino 93
Creating a BI Dashboard with Apache Superset 99
Summary 108

TABLE OF CONTENTS

Chapter 5: ETL Orchestration with Apache Airflow 109
Understanding Apache Airflow 110
    Airflow's Architecture and Components 111
    DAGs and Airflow Concepts 112
Starting Up the Lakehouse Components 113
Setting Up Apache Airflow Components 114
Configuring TLS and HTTPS for Trino 118
    Configuring Trino Connection in Airflow 122
Breaking Down the Airflow DAG File 124
    Connecting the DAG to Trino with Trino Operator 125
    Exporting Customer Segments into MinIO as CSV 128
    Notifying the Success of the Operation 128
Triggering the DAG 130
Verifying the Results of the DAG Execution 132
Cleaning Up the Environment 133
Summary 134

Part II: Streaming Data and Real-Time Analytics 137

Chapter 6: Real-time Change Data Capture with Kafka and Debezium 139
Introduction 139
Streaming Data, Apache Kafka, and Kafka Connect 140
Change Data Capture and Debezium 142
Before You Begin 142
Breakdown of the Docker-Compose File 143
Running the Setup 147
Source Database and Schema 147
Change Feed Creation 148

TABLE OF CONTENTS

Transforming the Change Event Structure 153

Sinking Change Events into OpenSearch 157

Verify End-to-End CDC Pipeline 160

Summary... 162

Chapter 7: Low-Latency Real-time Analytics Dashboard with ClickHouse 165

Introduction.. 165

Flash Sale Performance Dashboard 166

What Is Real-Time Analytics?... 166

    Business Value Over Batch Analytics... 167

    Key Components of Real-time Analytics Systems 167

Before You Begin... 168

Breakdown of the Docker-Compose File... 169

Running the Setup 172

Source Database and Schema 173

Creating the Change Data Feed for the purchases Table 174

    Using the Kafka Console Consumer 176

    Using the Redpanda Console 176

Configuring Clickhouse 177

Running the Streamlit Dashboard 181

Summary... 185

Chapter 8: Streaming ETL and Anomaly Detection with Apache Flink 187

Introduction.. 187

Login Anomaly Detection System 188

Stateful Stream Processing and Apache Flink 188

Before You Begin... 190

TABLE OF CONTENTS

Breakdown of the Docker-Compose File 190
Running the Setup 192
Creating Kafka Topics 193
Creating Flink SQL Tables 194
Simulating Login Events and Verification 201
Summary 204

Part III: Machine Learning and Feature Engineering 205

Chapter 9: Building a Product Recommendation Engine with Spark MLlib 207

Before You Begin 208
Setting Up the Lakehouse Components 209
Feature Engineering Pipeline 211
ALS Model Training Pipeline 216
Serving Recommendations with Flask 221
Summary 224

Chapter 10: Vector Similarity Search with Postgres and pgvector 227

Introduction 227
Introduction to Vector Embeddings and Similarity Search 229
    Understanding Vector Embeddings 229
    How Similarity Search Works 230
    Vector Databases: The Infrastructure for AI Applications 230
    PostgreSQL and pgvector: Democratizing Vector Search 231
Before You Begin 232
Start the Docker Setup 232
Configuring Postgres and pgvector 233
    Insert Sample Reviews 234

Search Frontend with Streamlit 235
    Vector Embeddings Generation 237
    Vector Similarity Search 239
Verify Semantic Search 242
Summary 243

Index 245

About the Author

Dunith Danushka has over 10 years of experience in the data analytics domain, including real-time analytics, stream processing, streaming data platforms, and event-driven systems. Throughout his career, he has developed, supported, and designed large-scale data-intensive applications, led teams who did that, and helped customers build their own.

His transition from engineering to solution architecture, developer relations, and now product marketing at EDB has given him a comprehensive outlook on the data space. His goal is to provide practical, hands-on guidance to help aspiring data engineers bridge the gap between theoretical knowledge and real-world applications.

About the Technical Reviewer

Shiroshica Kulatilake has over two decades of experience in the software industry, having worked across engineering, solutions architecture, and growth hacking, playing multiple roles. She has expertise in architecting and building both on-premises and cloud-based software and has designed and delivered enterprise integration solutions, event-driven systems, and product-led growth initiatives.

Shiroshica actively pursues data science and enjoys experimenting with new technologies. She values clarity, correctness, and practical applicability in technical content.

Acknowledgments

I would like to express my deepest gratitude to everyone who has influenced my journey in the data engineering space. This book would not have been possible without the collective wisdom shared by countless individuals throughout my career.

First and foremost, I'm thankful to all the colleagues and mentors I've had the privilege to work with over the past decade. Your patience, guidance, and willingness to share knowledge have been instrumental in shaping my understanding of complex data systems.

I'm also indebted to the broader data engineering community – the countless authors, bloggers, YouTubers, and conference speakers who generously share their expertise. Special thanks to the maintainers and contributors of the Apache projects featured in this book. Your dedication to open-source software has democratized access to powerful data tools and enabled countless innovations.

The online learning platforms, tutorial creators, Stack Overflow contributors, and GitHub project maintainers deserve special recognition. Your commitment to education and problem-solving has created an invaluable resource for practitioners at all levels.

To my family and friends who supported me through this writing process – thank you for your patience and encouragement. Writing a technical book while maintaining a full-time career is no small feat, and your support made it possible.

Finally, I want to acknowledge you, the reader. Your desire to learn and grow in the field of data engineering is what makes resources like this book meaningful. I hope that this work helps you on your journey, just as others have helped me on mine.

ACKNOWLEDGMENTS

This book represents my attempt to give back to the community that has given me so much. If it helps even one person navigate the complex world of data engineering more effectively, then it will have fulfilled its purpose.

Introduction

Welcome to *Practical Data Engineering with Apache Projects*, a comprehensive guide designed to bridge the gap between theoretical data engineering concepts and real-world applications. This book takes a hands-on approach to solving everyday data challenges using popular Apache projects like Spark, Iceberg, Kafka, Flink, and more.

In today's data-driven world, organizations of all sizes face complex challenges in collecting, processing, storing, and analyzing vast amounts of data. While there are many resources available that explain individual data technologies, there's a noticeable gap in the literature that demonstrates how these technologies work together to solve real business problems. This book aims to fill that gap.

Who This Book Is For

This book is primarily written for

- Aspiring data engineers who want to gain practical experience beyond theoretical concepts
- Software engineers transitioning into data engineering roles
- Data professionals looking to expand their toolkit with Apache projects
- Technical leaders evaluating data engineering solutions for their organizations
- Students and educators seeking real-world data engineering examples

INTRODUCTION

While some familiarity with programming (particularly Python) and basic data concepts is assumed, each chapter builds the necessary knowledge from the ground up, making the content accessible to readers with various levels of experience.

Book Structure

The book is organized into three main parts, each focusing on different aspects of data engineering:

Part 1: Data Lakehouses, Iceberg, Batch ETL, and Orchestration - This part focuses on building data storage solutions and implementing batch processing pipelines. You'll learn how to set up a data lakehouse with Apache Iceberg, create ETL pipelines with Spark, visualize data with Superset, and orchestrate workflows with Airflow.

Part 2: Streaming Data and Real-Time Analytics - This part explores real-time data processing using technologies like Kafka, Debezium, and Flink. You'll implement change data capture, streaming ETL, fraud detection, and a low-latency analytics dashboard with ClickHouse.

Part 3: Machine Learning and Feature Engineering - The final part delves into advanced applications, including feature engineering for machine learning and vector similarity search for sentiment analysis.

A Practical Approach

What sets this book apart is its practical, use-case-driven approach. Each chapter

- Presents a real-world business problem faced by different departments (IT, Marketing, Customer Support)
- Explains the relevant concepts and technologies needed to solve the problem

- Provides step-by-step instructions to implement a working solution
- Discusses how the solution relates to broader data engineering principles

All examples are based on a fictional e-commerce company, allowing you to see how different data engineering solutions work together in a cohesive business context.

Getting the Most from This Book

To maximize your learning experience:

- Follow along with the provided code examples (available in the companion GitHub repository)
- Experiment with the Docker-based environments that accompany each chapter
- Complete the exercises at the end of each chapter to reinforce your understanding
- Apply the concepts to your own projects or workplace challenges

By the end of this book, you'll have gained practical experience with a comprehensive stack of Apache data engineering tools and be well-equipped to implement similar solutions in your professional work.

Let's begin our journey into practical data engineering!

PART I

Data Lakehouses, Iceberg, Batch ETL, and Orchestration

CHAPTER 1

Foundational Data Engineering Concepts

What Is Data Engineering?

Data engineering is about building systems that take raw data from various sources, transform it into meaningful information, and make it accessible for others to use.

This is a broad definition of data engineering. As someone working in or aspiring to enter the field, you likely want a more technical explanation. Let's establish some context before diving into a more detailed definition.

Operational Systems vs. Analytical Systems

Organizations today generate and collect data through their day-to-day operations. This is what we call operational data that comes from various enterprise applications. For example, transaction records from sales systems, customer information in CRM systems, inventory management data, employee records in HR systems, and many more.

These operational systems are designed to handle many concurrent users and frequent updates, focusing on recording individual transactions and maintaining the current state. They're optimized for fast, reliable processing of day-to-day business operations, not for complex analysis.

CHAPTER 1   FOUNDATIONAL DATA ENGINEERING CONCEPTS

In contrast, analytical systems take this operational data and transform it into a format that's optimized for analysis and reporting. While operational systems handle the "now," analytical systems help understand trends, patterns, and insights over time.

The problem we often face is extracting this operational data from business systems and moving it to analytical systems. This is challenging because enterprise applications store data in their own, proprietary formats, resulting in *data silos*. These data silos are typically owned and controlled by individual departments or business units, spread over different geographic locations.

This is where data engineering comes in. Data engineers build pipelines that connect silos and transform data for analysis. This is commonly done through two approaches:

- **Extract, Transform, Load (ETL):** Data is extracted from source systems, transformed, and then loaded into analytical systems. This traditional approach is ideal when data requires significant cleansing or transformation before analysis.

- **Extract, Load, Transform (ELT):** Data is extracted, loaded into the target system first, and then transformed. This modern approach leverages the processing power of analytical systems and works well with cloud data warehouses.

For the purpose of this book, we'll primarily focus on the ETL approach, though we'll acknowledge where ELT might be more appropriate. Both methods enable organizations to develop an *enterprise view* across different data sources, making it possible to analyze data efficiently across the entire organization.

Role of a Data Engineer

A Data Engineer is a technology professional who designs, builds, and maintains the data infrastructure that enables organizations to collect, store, process, and analyze large volumes of data efficiently. They bridge the gap between operational systems and analytical systems by creating robust data pipelines and ensuring data quality and accessibility.

Key responsibilities of a data engineer include

- **Building and maintaining data pipelines:** Developing automated workflows to extract data from various sources, transform it according to business rules, and load it into target systems.

- **Data architecture design:** Creating scalable data storage solutions and implementing appropriate data models that support both operational and analytical needs.

- **Data quality management:** Implementing validation checks, monitoring data quality, and ensuring data consistency across different systems.

- **Performance optimization:** Tuning database performance, optimizing queries, and ensuring efficient data processing.

- **Data security and governance:** Implementing security controls, managing access permissions, and ensuring compliance with data protection regulations.

- **Infrastructure management:** Maintaining data infrastructure, handling system upgrades, and ensuring high availability of data services.

CHAPTER 1   FOUNDATIONAL DATA ENGINEERING CONCEPTS

Data engineers work closely with data scientists, analysts, and other stakeholders to ensure that data is available in the right format, at the right time, and with the right level of quality for various analytical needs.

As a data engineer, you need proficiency in various tools and scripting languages – especially SQL and Python.

- **SQL** – SQL (Structured Query Language) is essential for data engineers. This straightforward language lets you work directly with data in tables through commands like SELECT, INSERT, UPDATE, and DELETE.

- **Python** – Python ranks among the world's most popular programming languages. While versatile enough for web development and data analysis, it's particularly valuable in data engineering for its notebook-based workflows and machine-learning capabilities.

- **Others** – Your toolkit may expand to include languages like R, Java, Scala, or .NET, depending on your organization's needs and your expertise. Modern notebook environments support these languages, enabling seamless collaboration across different programming tools.

The Data Engineering Lifecycle

The data engineering lifecycle is an important process that helps build better data systems. Think of it as a step-by-step guide that shows how data moves from its starting point to where it's needed. This process makes sure the data stays accurate and reliable the whole way through.

Each step in this lifecycle has a specific job. When these steps work together properly, they create a strong system that helps businesses use their data effectively.

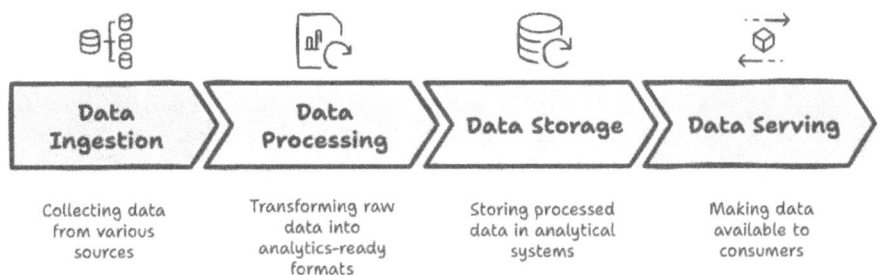

Figure 1-1. The data engineering lifecycle

Data Ingestion

The first phase is Data Ingestion, which involves collecting data from various data sources, the operational systems we discussed earlier. This phase can utilize batch ingestion to process data at scheduled intervals, stream ingestion to process data in real-time, or hybrid approaches that combine both methods.

During ingestion, engineers must carefully manage data validation and quality checks, handle schema evolution, minimize the impact on source systems, and implement proper rate limiting and back pressure handling mechanisms to ensure smooth data flow.

Data Processing

Data Processing is the second phase, where raw data is transformed into analytics-ready formats for analytical systems. This involves cleaning and standardizing data, enriching it with context, creating aggregations, and preparing features for machine learning applications.

Engineers can use various processing strategies like ETL, ELT (Extract, Load, Transform), stream processing, or hybrid approaches based on specific needs.

Data Storage

Data Storage is the third phase, where processed data is stored in analytical systems. The primary goals of this phase are to ensure cost-efficient storage and fast data retrieval. Data should be stored in analytics-ready columnar formats to optimize query performance and reduce storage costs.

Storage solutions can be implemented either on-premises or in the cloud, with cloud solutions offering advantages like infinite scalability, enhanced reliability, and robust security features. This phase focuses on implementing appropriate storage solutions for different data types and use cases, including data warehouses for structured data, data lakes for raw data storage, and lake houses that combine both approaches.

Engineers must carefully consider various factors such as data format selection (particularly columnar formats like Parquet, Avro, or ORC), partitioning strategies, compression techniques, and optimization of access patterns to ensure both efficient storage and quick data retrieval for downstream consumers.

Data Serving

Data Serving is the final phase where processed data is made available to consumers from analytical systems. Different consumers have varying expectations for data formats and use different methodologies to access the data.

- **Data scientists and analysts:** Utilize SQL interfaces for direct querying of data warehouses, work with Jupyter Notebooks or Google Colab for exploratory analysis, and leverage specialized data formats like Parquet or Arrow for efficient data processing.

- **Business intelligence teams:** Teams work with pre-aggregated data views for faster dashboard rendering, utilize OLAP cubes for multi-dimensional analysis, and establish direct connections to BI tools like Tableau or Power BI.

- **Applications and microservices:** These systems use REST APIs with appropriate authentication, GraphQL endpoints for flexible data querying, and message queues for event-driven architectures.

- **Business stakeholders:** Stakeholders access data through automated report generation in PDF or Excel formats, email-based data delivery systems, and self-service analytics platforms.

- **External partners:** Partners interact with data through secure FTP transfers for batch exports, API access with rate limiting and monitoring, and data sharing agreements through cloud storage.

CHAPTER 1 FOUNDATIONAL DATA ENGINEERING CONCEPTS

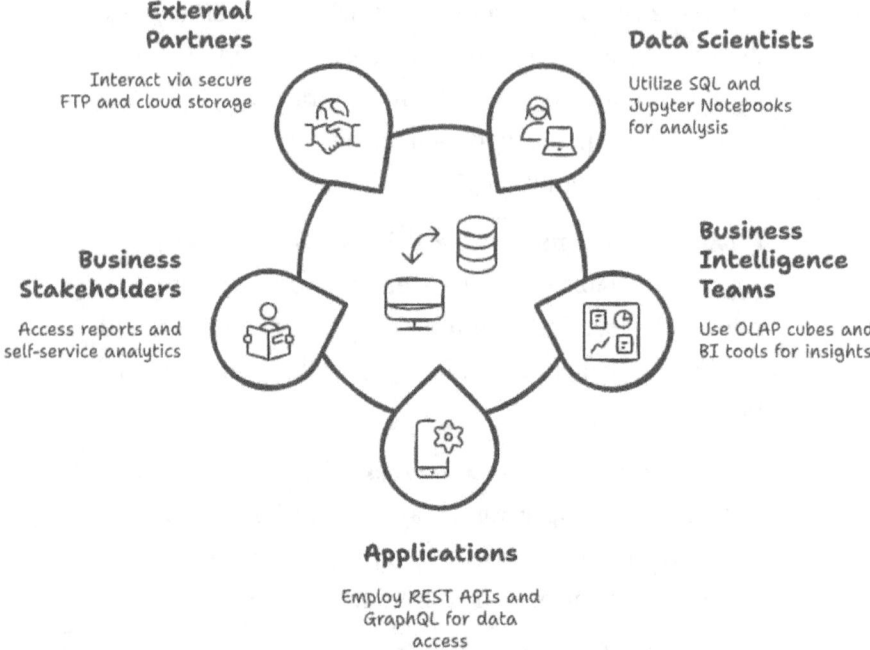

Figure 1-2. Processed data has many consumers

Each methodology is implemented with appropriate security controls, monitoring, and SLAs to ensure reliable data delivery while maintaining data governance standards.

The Apache Software Foundation and Data Engineering

The Apache Software Foundation (ASF) has played a pivotal role in shaping modern data engineering, though its origins were much broader in scope. Founded in 1999 around the Apache HTTP Server project, the foundation initially focused on general software engineering needs and web infrastructure. Over the decades, it has evolved dramatically to address emerging technological challenges, expanding into big data, cloud computing, and distributed systems. Today, it stands as one of the world's largest open-source software organizations, with its impact on data engineering being particularly profound. The ASF now hosts and maintains many of the most critical tools and frameworks that form the backbone of modern data infrastructure and processing pipelines. Apache has made a huge impact on how we work with big data. They've helped create many important tools that have changed how we handle large amounts of data. The way Apache works is special – they let anyone contribute based on their skills and experience, not their job title or position. This open approach, called "The Apache Way," encourages people to work together, share ideas openly, and make software that's free for everyone to use. Because of this approach, they've created many reliable and useful software tools that are now used by companies worldwide.

Apache's Contribution to Big Data Ecosystem

Apache Spark and Kafka have become fundamental technologies in data engineering. Spark transformed big data processing through its unified analytics engine, while Kafka established itself as the gold standard for distributed event streaming. Apache Iceberg has emerged as a key table format for managing large analytic datasets. We'll explore several other important Apache projects: NiFi for data ingestion, Flink for stream processing, Airflow for workflow orchestration, and Superset for data visualization – each serving a distinct role in the data engineering lifecycle.

The following chapters take a hands-on approach to these Apache projects. Instead of providing shallow coverage of many tools, we'll focus on mastering key technologies that address specific data engineering challenges. Each chapter offers detailed implementations, best practices, and real-world considerations for building production-grade data solutions.

CHAPTER 1   FOUNDATIONAL DATA ENGINEERING CONCEPTS

Introducing OneShop – Case Study for the Book

Throughout this book, we'll explore real-world data engineering concepts through the lens of a practical case study. Instead of abstract examples, we'll focus on solving concrete business challenges faced by OneShop, an e-commerce platform. By examining how different Apache tools and technologies can address OneShop's specific needs, you'll gain hands-on experience in implementing data solutions that drive business value. Each chapter will tackle different aspects of OneShop's data requirements, from processing customer transactions to building a conversational AI agent, demonstrating how various Apache projects work together in a production environment.

OneShop is a bustling online e-commerce platform that offers more than 1,000 unique items across various categories. From trending gadgets to household essentials, OneShop serves a wide range of customer needs. Over the years, the store has grown significantly, serving more than 1,000 page views per minute on its website and processing around 50 orders during peak times.

OneShop currently operates three systems that handle day-to-day activities:

1. **PostgreSQL Database**: This relational database houses key business data, including:

    - **Users table**: Information about customers who shop on the platform.

    - **Items table**: A catalog of the more than 1,000 products available for sale.

    - **Purchases table**: Records of transactions, capturing details about what customers buy and when.

CHAPTER 1 FOUNDATIONAL DATA ENGINEERING CONCEPTS

2. **Store website**: The primary customer touchpoint, where visitors browse items, add them to their cart, and complete purchases. The website generates a wealth of data, including user behavior, product views, and click patterns.

3. **Data lake**: Built on MinIO, the data lake serves as a repository for storing raw, semi-structured, and unstructured data. Currently, it collects page view events from the website, tracking every product a user explores during their visit.

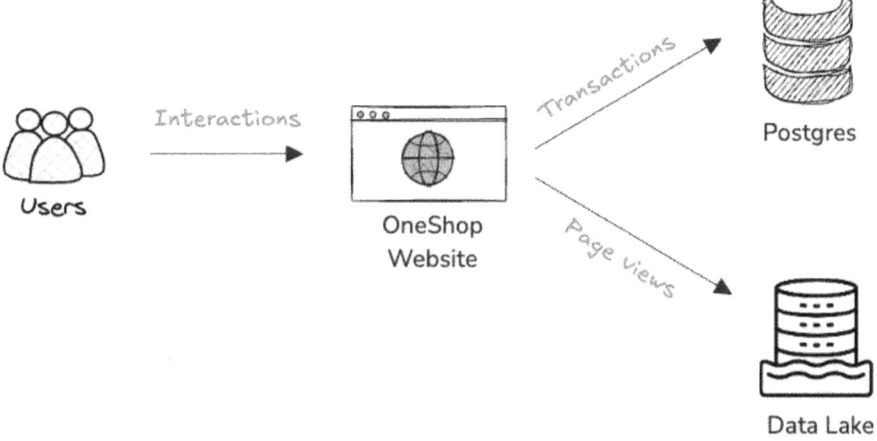

Figure 1-3. OneShop's current architecture

These systems form the backbone of OneShop's operations, but they're just the beginning. The OneShop management team plans to leverage data analytics and AI advancements to boost revenue while maintaining excellent customer service. As OneShop continues to scale, the platform requires a comprehensive data engineering strategy to manage both current and future data demands.

CHAPTER 1 FOUNDATIONAL DATA ENGINEERING CONCEPTS

This strategy must address the unique needs and challenges of three key departments: IT, Marketing, and Customer Support. The IT department needs reliable, accurate, and timely analytics, while Marketing seeks deeper insights into customer behavior and campaign effectiveness. The Customer Support team requires tools for customer sentiment analysis and improved service delivery. To meet these requirements, OneShop has launched a data engineering program to transform its infrastructure into a robust, scalable platform that delivers actionable insights and enhances the shopping experience.

As the new data engineer at OneShop, you will work closely with stakeholders from each department to understand their needs and develop data pipelines that bring value to their operations. Throughout this book, you'll implement ten practical data engineering pipelines, each addressing a unique departmental use case.

Roadmap

This book is organized into three distinct parts, each focusing on different aspects of modern data engineering:

Part 1: Foundational Data Engineering. The first part of the book covers essential data engineering concepts and tools, including data storage in lakehouses, ETL pipelines, data visualization, and pipeline automation.

- **Chapter 2** – You will set up an Apache Iceberg data lakehouse infrastructure from the ground up. This lakehouse provides a strong foundation for the OneShop data engineering team. You will see how they build several projects based on this infrastructure.

CHAPTER 1   FOUNDATIONAL DATA ENGINEERING CONCEPTS

- **Chapter 3** – You will model the Iceberg lakehouse (set up in Chapter 2) with Medallion architecture. You'll develop batch ETL pipelines with Apache Spark to load data into the bronze layer and transform this data for the silver layer.

- **Chapter 4** – You will define gold layer tables in the lakehouse using Trino. Then, you will use Apache Superset to create business intelligence dashboards from these gold tables.

- **Chapter 5** – You will use Apache Airflow to orchestrate an ETL pipeline that computes customer segments based on silver tables in the lakehouse, exports the results to MinIO as a CSV file, and sends an email reminder.

Part 2: Streaming and Real-Time Analytics. The second part focuses on handling real-time data and stream processing with the Apache Kafka ecosystem. You'll explore:

- **Chapter 6** – You will implement a change data capture (CDC) pipeline with Debezium to capture inventory level changes from Postgres and reliably update a search index in OpenSearch.

- **Chapter 7** – You will develop a real-time analytics dashboard with Apache Kafka, Clickhouse, and Streamlit to visualize OneShop's flash sale campaigns in real-time.

- **Chapter 8** – You will use Kafka and Apache Flink to build a user login anomaly detection system.

Part 3: Data Engineering for AI and ML. The final part demonstrates how data engineering enables modern AI and machine learning applications. Projects include

- **Chapter 9** – You will build a product recommendation engine for OneShop. We will use Spark MLlib library to create a machine learning feature engineering pipeline based on the data available in the silver layer tables, computing the features required for the recommender model, and storing the refined features in a gold layer table.

- **Chapter 10** – As the final project, you will implement a semantic similarity search engine to analyze customer reviews left by OneShop customers. You will use the pgvector extension on Postgres for this.

Each part builds upon the previous sections, providing you with a comprehensive understanding of modern data engineering practices and tools. The hands-on projects will give you practical experience in implementing real-world solutions using industry-standard Apache technologies.

Prerequisites

Before diving into the practical projects in this book, there are a few prerequisites you should have in place to ensure a smooth learning experience:

Docker Environment

Almost all the projects you will explore in this book are available as Docker Compose projects. This approach offers several benefits:

- **Self-contained environments:** Each project runs in isolated containers, preventing conflicts between different technology stacks.

- **Easy reproducibility:** You can quickly spin up complex multi-service architectures with a single command.

- **Consistent experience:** The containerized setup ensures that the projects work the same way across different operating systems and environments.

- **Simplified cleanup:** When you're done with a project, you can remove all associated resources without affecting your local system.

To run these projects, you will need to have the following installed on your local machine:

- **Docker engine** (version 20.10 or higher)
- **Docker compose** (version 2.0 or higher)

For the resource requirements, we recommend:

- **CPU:** At least 4 cores (8 cores recommended for smoother performance)
- **Memory:** Minimum 8GB RAM (16GB or more recommended, especially for projects involving Apache Spark)
- **Storage:** At least 20GB of free disk space

Project Code Repository

All code for the projects discussed in this book is available in the accompanying GitHub repository.

CHAPTER 1   FOUNDATIONAL DATA ENGINEERING CONCEPTS

To clone the repository using a Git client:

https://github.com/Apress/Practical-Data-Engineering-with-Apache-Projects

Alternatively, you can download the code as a ZIP file.

The repository is structured to optimize your learning experience, with separate folders organized by chapter.

```
.
├── chapter-02
├── chapter-03
├── chapter-04
├── chapter-05
├── chapter-06
├── chapter-07
├── chapter-08
├── chapter-09
├── chapter-10
```

To make your learning experience smoother, we've pre-implemented the difficult or boilerplate parts of each project, allowing you to focus on building the core data engineering components while following along with each chapter. This approach saves you valuable time by letting you concentrate on the concepts that matter most. Additionally, each project comes with clear instructions on how to get started and which components you need to implement.

Programming Knowledge

To get the most out of this book, you should have familiarity with the following technologies and languages:

CHAPTER 1   FOUNDATIONAL DATA ENGINEERING CONCEPTS

- **Python:** Most examples use Python for data processing, transformation, and pipeline development.
- **PySpark:** Basic knowledge of PySpark APIs will be helpful for working with large-scale data processing.
- **Java:** Some components, particularly those related to Apache Kafka and Flink, use Java.
- **YAML:** Configuration files for Docker Compose and various services are written in YAML.
- **SQL:** You should be comfortable with basic to intermediate SQL queries for data analysis and transformation.

If you need to brush up on any of these skills, we recommend spending some time refreshing your knowledge before diving into the more complex projects. There are many online resources available for quick reviews of these technologies.

Summary

In this introductory chapter, we established the foundation for understanding modern data engineering. We defined data engineering as the practice of building systems that transform raw data into meaningful, accessible information. We explored the fundamental phases of the data engineering lifecycle: ingestion, processing, storage, and serving, highlighting how each phase contributes to creating value from data.

The chapter presented OneShop, our case study e-commerce platform, which will serve as our practical example throughout the book. We examined OneShop's current architecture, including its PostgreSQL database, website, and MinIO data lake. We outlined the challenges faced by different departments (IT, Marketing, and Customer Support) and how data engineering solutions can address these needs.

Finally, we covered the prerequisites needed to follow along with the practical examples, including Docker environment requirements, access to the project code repository, and recommended programming knowledge in Python, PySpark, Java, YAML, and SQL.

Throughout the upcoming chapters, we'll build hands-on experience with industry-standard Apache tools while addressing real business needs, providing you with practical skills applicable to real-world data engineering challenges.

CHAPTER 2

Implementing the Data Lakehouse

As OneShop continues to grow, the engineering team recognizes the need for a more sophisticated data management approach. The increasing data volume from website interactions, the growing complexity of user behavior analysis, and the desire to implement advanced analytics and AI capabilities have pushed them to explore modern data architectures.

The team has identified several key requirements: they need a system that can handle both structured data (like user and purchase records from MySQL) and semi-structured data (like website interaction events) while maintaining data consistency and enabling efficient queries. Additionally, they want an architecture that can support future machine learning and AI initiatives without requiring significant infrastructure changes.

The team has identified several key requirements:

- **Data integration challenges:** Combine data from multiple source systems, including their e-commerce platform, marketing tools, and customer service systems, into a unified platform.

- **Scalability requirements:** Build a solution that can handle their rapidly growing data volume while maintaining query performance and cost efficiency.

- **Analytics modernization:** Create a foundation for advanced analytics capabilities, including real-time dashboards, machine learning, and Generative AI.

After careful evaluation, OneShop's engineering team has decided to implement a data lakehouse architecture using Apache Iceberg as the table format. This approach will allow them to combine the flexibility of their existing data lake with the reliability and performance traditionally associated with data warehouses. In this chapter, we'll walk through the process of setting up this lakehouse architecture, demonstrating how it addresses OneShop's current needs while preparing them for future data initiatives.

OneShop Data Lakehouse

As a newly hired data engineer at OneShop, you've been tasked with implementing the data lakehouse using open-source components to minimize initial costs. As a beginner to data lakehouses and Apache Iceberg, you face several challenges:

- Understanding how different components of a data lakehouse work together
- Learning Iceberg's features through hands-on practice
- Selecting appropriate open-source data infrastructure technologies for each lakehouse component
- Setting up a local development environment for Iceberg-based projects

This chapter addresses these challenges by providing a complete local development environment using Docker. You'll explore Apache Iceberg's features and experiment with different configurations before moving to a large-scale deployment. After reading this chapter, you will build the following minimal data lakehouse architecture that includes all essential components: Apache Spark for processing, MinIO for storage, and Iceberg REST catalog for metadata management.

Throughout this book, we'll expand upon this lakehouse architecture to handle more complex use cases. First, however, let's explore the core concepts of a data lakehouse and understand why it exists.

The Problems of Data Warehouses and Data Lakes

Traditional data warehouses, while providing reliable data management with ACID compliance, data consistency, optimized queries, and robust security, faced significant limitations that hindered their effectiveness. The combination of poor scalability, high storage costs from proprietary solutions, inflexible schemas, and the inability to handle unstructured data created a cascade of challenges: organizations were forced to maintain multiple systems, implement complex ETL processes, make substantial infrastructure investments, and manage data silos that ultimately hampered comprehensive analytics and made it prohibitively expensive to maintain historical data.

Organizations initially turned to data lakes to address these limitations. Data lakes offer cost-effective storage and flexibility for various data types, enabling organizations to unlock numerous use cases for semi-structured and unstructured data analytics. Data lakes became instrumental in processing streaming data from IoT devices, social media feeds, and real-time applications. Additionally, they provided the foundation for advanced machine learning and AI applications by allowing data scientists to work with large, diverse datasets in their raw form.

However, data lakes brought in their own set of problems. They often suffer from poor data quality due to a lack of schema enforcement as they typically employ a "schema on read" approach. They also lack proper

CHAPTER 2   IMPLEMENTING THE DATA LAKEHOUSE

metadata management, indexing, and query optimization techniques, resulting in a suboptimal query performance. Most importantly, data lakes lack a crucial feature of transactional guarantees, which means they cannot ensure data consistency and reliability during concurrent operations.

With these limiting factors, organizations often found themselves maintaining two separate systems to handle different data types and workloads effectively: data lakes for storing raw, unstructured data and supporting data science workloads, and data warehouses for structured data and business intelligence needs. A collection of ETL (Extract, Transform, Load) pipelines extracts data from the lake, transforms it to match warehouse schemas, and loads it into the warehouse.

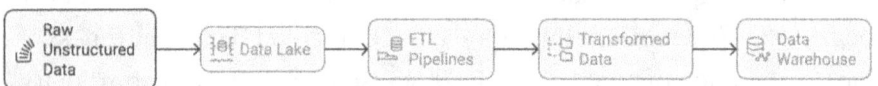

This dual-system approach led to several operational challenges. The maintenance of complex ETL pipelines became burdensome, requiring constant scheduling and monitoring. The setup also increased storage costs through data redundancy, created delays between data ingestion and analysis, and made it difficult to maintain consistency between systems.

This fragmented approach created significant delays in data availability for business insights.

What Is a Data Lakehouse?

A data lakehouse is a modern data management architecture that combines the best features of data lakes and data warehouses. It was introduced to bridge the gap between these two traditional architectures, offering both the flexibility of data lakes and the performance and reliability of data warehouses.

The key innovation of data lakehouses is their ability to enforce data structure and schema while maintaining the cost-effectiveness and scalability of object storage. This is achieved through specialized table formats and metadata management systems that provide ACID transactions, schema enforcement, and efficient query optimization – all while working directly with files stored in object storage.

The unified data lakehouse architecture eliminates the need for separate systems by allowing data science, machine learning, and business intelligence workloads to operate on a single platform. This approach enables effective management of both structured and unstructured data while reducing overall costs and complexity.

What Is a Table Format?

A key enabler of the data lakehouse architecture is the table format – a specification that defines how data files are organized, tracked, and managed to form a logical table. While file formats like Parquet and ORC handle how data is encoded within individual files, table formats operate at a higher level, managing collections of files as tables with features like schema evolution, partitioning, and transaction management. These table formats provide the crucial metadata layer that enables database-like capabilities on top of simple object storage systems, effectively bridging the gap between data lakes and data warehouses.

Several table formats have emerged to address these needs, including Apache Hudi, Delta Lake, and Apache Iceberg. Among these, Apache Iceberg stands out for its robust architecture and feature set. Developed at Netflix and later donated to the Apache Software Foundation, Iceberg provides a table format that ensures reliable, scalable data management with features like schema evolution, hidden partitioning, time travel, and ACID compliance, making it particularly well-suited for large-scale data lake implementations.

Now that we've got the definitions out of the way, it's time to kick off our first hands-on project.

Before You Begin

As we discussed in the **Prerequisites** section of the previous chapter, we have a companion GitHub repository containing the code samples that we are going to try out in the book. If you haven't done so, clone the repository to your local machine by running:

```
git clone https://github.com/Apress/Practical-Data-Engineering-with-Apache-Projects.git
```

You will find the code for this chapter located in the `chapter-02` folder. If you haven't set everything up yet, refer to the **Prerequisites** section in the previous chapter for more information.

Navigate to the project folder on a terminal by typing:

```
cd <repository_root>/chapter-02
```

In this folder, you will find everything you need to deploy a minimal data lakehouse on Docker Compose.

Breakdown of the Docker-Compose File

The `chapter-02/docker-compose.yaml` file defines services that create our data lakehouse environment.

```
version: "3"

services:
  spark-iceberg:
```

```
    container_name: spark-iceberg
    build: spark/
    networks:
      iceberg_net:
    depends_on:
      -rest
      -minio
    volumes:
      -./warehouse:/home/iceberg/warehouse
      -./notebooks:/home/iceberg/notebooks/notebooks
    environment:
      -AWS_ACCESS_KEY_ID=admin
      -AWS_SECRET_ACCESS_KEY=password
      -AWS_REGION=us-east-1
    ports:
      -8888:8888
      -8080:8080
      -10000:10000
      -10001:10001
rest:
    image: apache/iceberg-rest-fixture
    container_name: iceberg-rest
    networks:
      iceberg_net:
    ports:
      -8181:8181
    environment:
      -AWS_ACCESS_KEY_ID=admin
      -AWS_SECRET_ACCESS_KEY=password
      -AWS_REGION=us-east-1
      -CATALOG_WAREHOUSE=s3://warehouse/
```

```yaml
      -CATALOG_IO__IMPL=org.apache.iceberg.aws.s3.S3FileIO
      -CATALOG_S3_ENDPOINT=http://minio:9000
  minio:
    image: minio/minio
    container_name: minio
    environment:
      -MINIO_ROOT_USER=admin
      -MINIO_ROOT_PASSWORD=password
      -MINIO_DOMAIN=minio
    networks:
      iceberg_net:
        aliases:
          -warehouse.minio
    ports:
      -9001:9001
      -9000:9000
    command: ["server", "/data", "--console-address", ":9001"]
  mc:
    depends_on:
      -minio
    image: minio/mc
    container_name: mc
    networks:
      iceberg_net:
    environment:
      -AWS_ACCESS_KEY_ID=admin
      -AWS_SECRET_ACCESS_KEY=password
      -AWS_REGION=us-east-1
    entrypoint: |
      /bin/sh -c "
```

```
                until (/usr/bin/mc config host add minio <http://
                minio:9000> admin password) do echo '...waiting...' &&
                sleep 1; done;
                /usr/bin/mc rm -r --force minio/warehouse;
                /usr/bin/mc mb minio/warehouse;
                /usr/bin/mc policy set public minio/warehouse;
                tail -f /dev/null
                "
networks:
  iceberg_net:
```

Let's break down each service a little bit further.

- spark-iceberg – This is a custom Docker image with Python 3, Apache Spark with Iceberg support, a Jupyter Notebooks server, and PyIceberg SDK. This container spins up a Spark cluster in local mode, preconfigured with Iceberg runtime support.
- rest – Provides the Iceberg REST catalog for metadata management, using MinIO as its storage backend
- minio – Deploys the MinIO storage server
- mc – MinIO Client providing a CLI

If you already have Spark installed, you can add Iceberg support in two ways: use the `--packages` option when starting a Spark/PySpark shell, or add the `iceberg-spark-runtime-<version>` to Spark's jars folder. The spark-iceberg service has already performed this for you to save time.

CHAPTER 2   IMPLEMENTING THE DATA LAKEHOUSE

Deconstructing the Spark-Iceberg Dockerfile

This `./chapter-02/spark` folder holds the necessary files for building the spark-iceberg image.

The `./chapter-02/spark/Dockerfile` contains all the instructions for preparing a Spark runtime with Iceberg support. In addition to that, it assembles essential tools for a better Iceberg developer experience – to make working with Iceberg easier. Let's highlight only the configurations that are relevant to us.

First, we start building a new image based on the Python 3.10 base image. After that, essential Unix system utilities are installed along with `openjdk-11-jdk,` which provides the Java runtime for Spark.

```
FROM python:3.10-bullseye

RUN apt-get update && \\
    apt-get install -y --no-install-recommends \\
      sudo \\
      curl \\
      vim \\
      unzip \\
      openjdk-11-jdk \\
      build-essential \\
      software-properties-common \\
      ssh && \\
    apt-get clean && \\
    rm -rf /var/lib/apt/lists/*
```

The `requirements.txt` file in the same folder lists the Python dependencies that we are going to need. This includes the Jupyter Notebooks server, PyIceberg libraries, etc.

```
jupyter==1.0.0
spylon-kernel==0.4.1
pyiceberg[pyarrow,duckdb,pandas]==0.7.1
jupysql==0.10.5
matplotlib==3.9.2
scipy==1.14.1
duckdb-engine==0.13.1
```

They are installed with pip

```
COPY requirements.txt .
RUN pip3 install -r requirements.txt
```

Next, we define several environment variables and download the Apache Spark distribution. Notice the Spark and Iceberg versions we use here.

```
# Optional env variables
ENV SPARK_HOME=${SPARK_HOME:-"/opt/spark"}
ENV PYTHONPATH=$SPARK_HOME/python:$SPARK_HOME/python/lib/
py4j-0.10.9.7-src.zip:$PYTHONPATH

WORKDIR ${SPARK_HOME}

ENV SPARK_VERSION=3.5.2
ENV SPARK_MAJOR_VERSION=3.5
ENV ICEBERG_VERSION=1.6.0

# Download spark
RUN mkdir -p ${SPARK_HOME} \\
 && curl <https://dlcdn.apache.org/spark/spark-${SPARK_VERSION}/spark-${SPARK_VERSION}-bin-hadoop3.tgz> -o spark-${SPARK_VERSION}-bin-hadoop3.tgz \\
 && tar xvzf spark-${SPARK_VERSION}-bin-hadoop3.tgz --directory /opt/spark --strip-components 1 \\
 && rm -rf spark-${SPARK_VERSION}-bin-hadoop3.tgz
```

Once the Spark distribution is downloaded and extracted, we continue to download the Iceberg runtime for Spark.

```
# Download iceberg spark runtime
RUN curl <https://repo1.maven.org/maven2/org/apache/iceberg/iceberg-spark-runtime-${SPARK_MAJOR_VERSION}_2.12/${ICEBERG_VERSION}/iceberg-spark-runtime-${SPARK_MAJOR_VERSION}_2.12-${ICEBERG_VERSION}.jar> -Lo /opt/spark/jars/iceberg-spark-runtime-${SPARK_MAJOR_VERSION}_2.12-${ICEBERG_VERSION}.jar
```

Notice how the Iceberg jars are placed inside the `jars` folder of the Spark installation, which is located in the `/opt/spark` folder of the image.

Next, we create the `notebook` command that enables the container to start a Jupyter Notebook server configured for PySpark by simply running it.

```
# Add a notebook command
RUN echo '#! /bin/sh' >> /bin/notebook \
 && echo 'export PYSPARK_DRIVER_PYTHON=jupyter-notebook' >> /bin/notebook \
 && echo "export PYSPARK_DRIVER_PYTHON_OPTS=\\"--notebook-dir=/home/iceberg/notebooks --ip='*' --NotebookApp.token='' --NotebookApp.password='' --port=8888 --no-browser --allow-root\\"" >> /bin/notebook \
 && echo "pyspark" >> /bin/notebook \
 && chmod u+x /bin/notebook
```

You will also find the `spark-defaults.conf` located in the same folder. The following line copies it to Spark's configuration folder.

```
COPY spark-defaults.conf /opt/spark/conf
```

CHAPTER 2   IMPLEMENTING THE DATA LAKEHOUSE

The `spark-defaults.conf` file is a configuration file used in Apache Spark to set default configuration parameters for Spark applications. It allows you to define system-wide default settings that will be applied to all Spark jobs running on a cluster, such as memory allocations, logging levels, performance tuning parameters, and other runtime configurations.

In the `spark-defaults.conf` file, you will notice several configurations instructing Spark to set up a REST-based Iceberg catalog (demo) using MinIO as the object storage backend.

```
spark.sql.catalog.demo                  org.apache.iceberg.
                                        spark.SparkCatalog
spark.sql.catalog.demo.type             rest
spark.sql.catalog.demo.uri              <http://rest:8181>
spark.sql.catalog.demo.io-impl          org.apache.iceberg.aws.
                                        s3.S3FileIO
spark.sql.catalog.demo.warehouse        s3://warehouse/wh/
spark.sql.catalog.demo.s3.endpoint      <http://minio:9000>
```

If you look closely, you will see that the `spark.sql.catalog.demo.uri` points to `http://rest:8181`, which is the REST catalog's endpoint deployed in the `rest` container, enabling Spark to communicate with the catalog for metadata operations.

`spark.sql.catalog.demo.s3.endpoint` points to the MinIO server container, allowing Iceberg to use MinIO as the storage backend for the catalog, while `spark.sql.catalog.demo.warehouse` defines the root directory where table data and metadata are stored in MinIO. Lastly, `spark.sql.catalog.demo.io-impl` configures Iceberg to use the S3FileIO implementation, enabling it to read from and write to S3-compatible object storage like MinIO.

Finally, we copy the `.pyiceberg.yaml` file to the root directory of the container. This file configures the PyIceberg runtime to work with the Iceberg REST catalog. We will discuss the contents of the file in detail later.

```
COPY .pyiceberg.yaml /root/.pyiceberg.yaml
```

While this Dockerfile is a long one, it simplifies many things for you that you'd have to do manually otherwise. Next, let's start everything up.

Running Everything

In the same directory as the `docker-compose.yaml` file, run the following command to build the spark-iceberg image.

`docker-compose build`

Then, start all the containers.

`docker-compose up -d`

The runtime provided by the docker-compose file is far from a large-scale production-grade warehouse, but it does let you explore Iceberg's wide range of features.

The following table lists these components with the URLs they expose:

Component	URL
Jupyter Notebook server	http://localhost:8888
Spark driver UI	http://localhost:8080
MinIO storage server	http://localhost:9001
Iceberg REST catalog	http://localhost:8181

CHAPTER 2   IMPLEMENTING THE DATA LAKEHOUSE

The runtime representation of the lakehouse architecture would look like this:

As shown in Figure 2-1, the minimal lakehouse architecture for OneShop consists of Apache Spark for data processing, a MinIO-based data lake for storage, and an Iceberg REST catalog for metadata management.

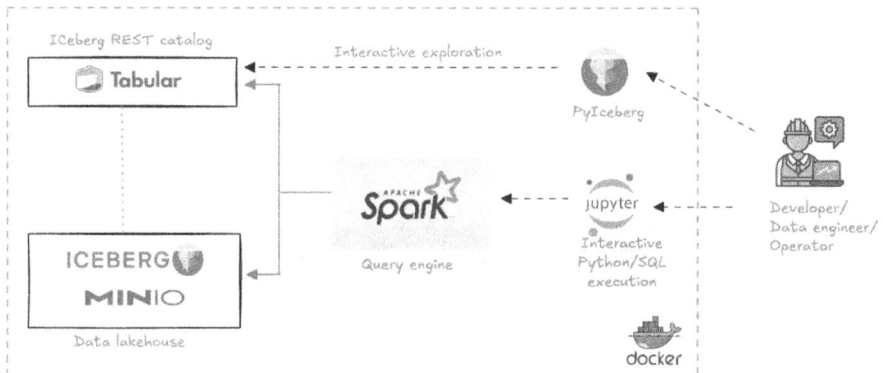

Figure 2-1. *Runtime representation of the lakehouse architecture*

It will take some time to get all the services for the first time. Once all the services are up and running, head over to localhost:9001 to log in to the MinIO console with the username "*admin*" and the password "*password*".

CHAPTER 2 IMPLEMENTING THE DATA LAKEHOUSE

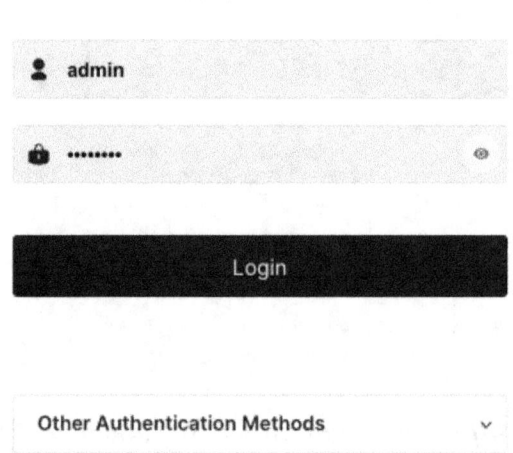

Figure 2-2. *MinIO Console login*

These credentials come from the environment variables defined in the docker-compose.yml file.

```
minio:
    image: minio/minio
    container_name: minio
    environment:
      -MINIO_ROOT_USER=admin
      -MINIO_ROOT_PASSWORD=password
```

Feel free to change them as you want.

Once you log in to the console, click on the "buckets" section from the menu on the left, and you will notice a bucket named "warehouse". Minio is an S3-compatible storage layer, so a bucket is essentially where you can save files in object storage solutions like S3 and Minio.

Object Browser

Name	Objects	Size	Access
warehouse	12	24.9 KiB	R/W

Figure 2-3. *Bucket*

If you're wondering how this bucket was created automatically, it happens through the entrypoint command defined under mc, the MinIO Client, in the docker-compose.yml file.

```
entrypoint: >
     /bin/sh -c "
     until (/usr/bin/mc config host add minio <http://
     minio:9000> admin password) do echo '...waiting...' &&
     sleep 1; done;
     /usr/bin/mc rm -r --force minio/warehouse;
     /usr/bin/mc mb minio/warehouse;
     /usr/bin/mc policy set public minio/warehouse;
     tail -f /dev/null
```

This command removes any existing content in the minio/warehouse bucket using the mc rm -r --force minio/warehouse command. This ensures that the bucket is empty before any new operations are performed. Next, it creates a new bucket named warehouse in the MinIO server using the mc mb minio/warehouse command. After creating the bucket, it sets the bucket's policy to public, making the contents of the bucket accessible to the public.

CHAPTER 2   IMPLEMENTING THE DATA LAKEHOUSE

Creating a Database and an Iceberg Table

Now that our development environment is up and running with all the necessary components – Spark, MinIO, and the REST catalog – we can start creating and working with Iceberg tables. Let's create a database and sample tables to explore Iceberg's features.

The first step is to access your development environment. In a browser, navigate to localhost:8888/lab to access the Jupyter Notebook server exposed by the `spark-iceberg` container.

You should see something like this:

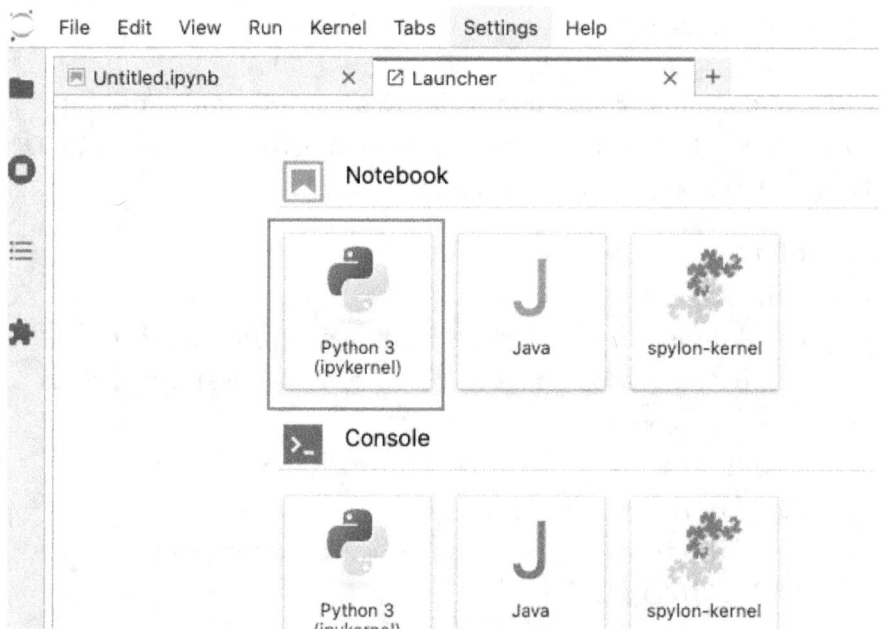

Figure 2-4. Jupyter notebook server workspace

Go ahead and create a new **Python 3** notebook as shown above.

While you can use the terminal to write code in the PySpark environment, we prefer Jupyter Notebooks because they offer an interactive workspace where you can write and test code immediately, while also allowing you to document your work with explanatory text and visualizations. The ability to organize code into separate cells, maintain a persistent record of your work, and easily create visual representations of your data makes notebooks particularly valuable for data analysis tasks.

Let's create a new database and a simple table that has one partition. Insert the following code in a new cell to create a new database named accounts.

```
%%sql
CREATE DATABASE IF NOT EXISTS accounts;
```

You will notice that %%sql in the first line, which is a Jupyter Notebook magic command that allows us to write SQL queries directly in a notebook cell. When we use this magic command, the cell's contents are interpreted as SQL rather than Python code. This is particularly useful when working with Spark SQL as it saves us from having to wrap our SQL queries in Python strings and calling Spark SQL functions explicitly. This magic command comes from JupySQL, which we installed via the requirements.txt file while building the spark-iceberg image.

```
jupysql==0.10.5
duckdb-engine==0.13.1
```

Next, place the following code block to create the users table with one partition.

```
%%sql
CREATE TABLE IF NOT EXISTS accounts.users (
    id INT,
    first_name VARCHAR(25),
    last_name VARCHAR(25),
    email VARCHAR(50)
)
USING iceberg
PARTITIONED BY (truncate(1, last_name))
```

CHAPTER 2   IMPLEMENTING THE DATA LAKEHOUSE

The `truncate(1, last_name)` in the `PARTITIONED BY` clause demonstrates Iceberg's powerful **Hidden Partitioning** feature. This feature enables advanced partitioning strategies without creating extra columns that make querying the table more complicated. In this case, we're using a partition transform to group records based on the first character of each person's `last_name`.

Let's examine what happens in MinIO before adding any records. After creating the table, Iceberg generates a metadata file – the first level in its data architecture. To view this, log in to the MinIO Console at localhost:9001. As defined in the Docker Compose file, all data is stored in the `warehouse` bucket. Inside this bucket, you'll find the metadata file for the `users` table. Notice how the REST Catalog uses both the database and table names in the file path.

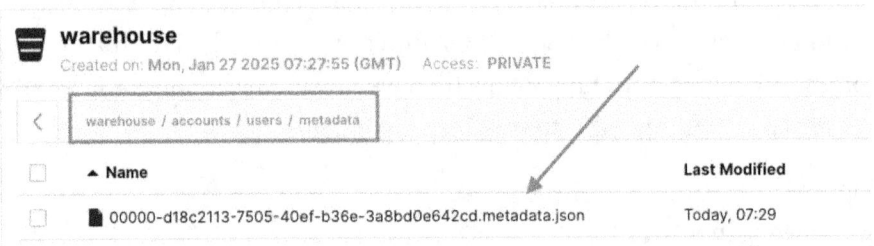

Figure 2-5. *Metadata file for the* `users` *table*

This is the only metadata file that exists in MinIO at this point since we have not added any data to the table.

While experimenting with Apache Iceberg, you may want to drop a table to start over. You can use the `DROP TABLE` clause to drop the `users` table if you need to modify its configuration.

```
%%sql
DROP TABLE IF EXISTS accounts.users;
```

40

Adding Records to the Table

Run the code below in a new notebook cell to add 10 records to the users table:

```
%%sql
INSERT INTO accounts.users (id, first_name, last_name, email) VALUES
(1, 'Juli', 'Arthars', 'jarthars0@mit.edu'),
(2, 'Matthiew', 'Hurley', 'mhurley1@narod.ru'),
(3, 'Lena', 'Westcarr', 'lwestcarr2@jimdo.com'),
(4, 'Reagen', 'Josifovitz', 'rjosifovitz3@hexun.com'),
(5, 'Ogden', 'Janecek', 'ojanecek4@yellowbook.com'),
(6, 'Calypso', 'McMurrugh', 'cmcmurrugh5@cam.ac.uk'),
(7, 'Dru', 'Garces', 'dgarces6@privacy.gov.au'),
(8, 'Jack', 'Matschoss', 'jmatschoss7@latimes.com'),
(9, 'Patty', 'Furnell', 'pfurnell8@yelp.com'),
(10, 'Lyndsie', 'Speeks', 'lspeeks9@skyrock.com');
```

After running this query, go back to MinIO and check the /warehouse/accounts/users/data/ directory. You'll find folders containing the partitioned files, each labeled according to its partition value.

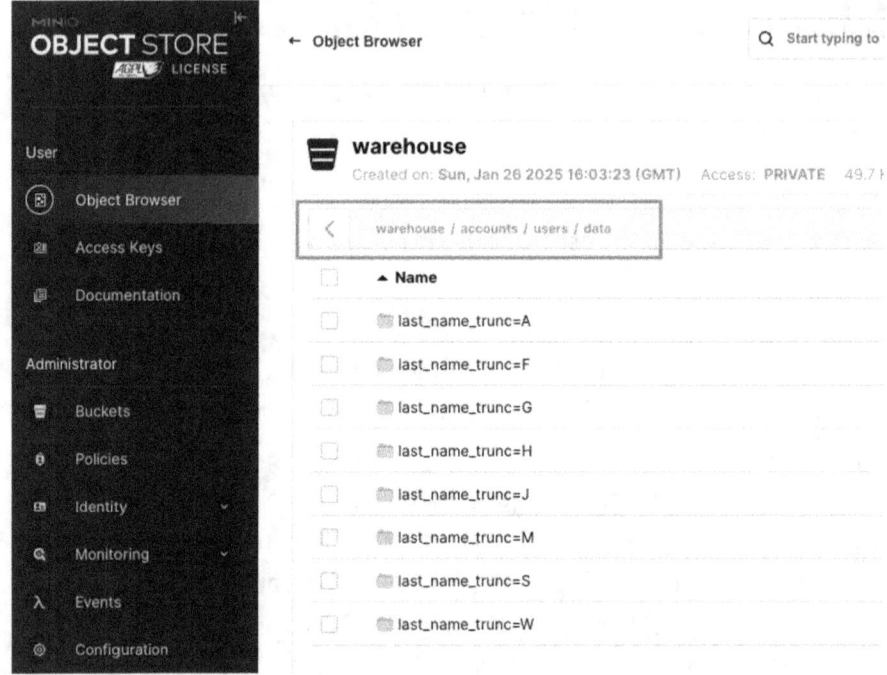

Figure 2-6. Partitioned files

Querying the Table with PyIceberg

While you can read all the data back from the users table with a simple SELECT query like this:

```
%%sql
SELECT * FROM accounts.users;
```

Let's try a more interesting approach using PyIceberg as it provides a pure Pythonic interface to interact with Iceberg tables, without requiring a Spark context or any JVM dependencies.

CHAPTER 2   IMPLEMENTING THE DATA LAKEHOUSE

If you recall the `requirements.txt` file, PyIceberg was preinstalled with `pyiceberg[pyarrow,duckdb,pandas]==0.7.1` during the creation of the `spark-iceberg` image.

Run the below code in a new notebook cell. It will return all rows from the `users` table.

```
from pyiceberg.catalog import load_catalog

# Load the REST catalog
catalog = load_catalog('default')

# Load the users table
tbl = catalog.load_table('accounts.users')

# Create a scan with filter on last_name
sc = tbl.scan(row_filter="last_name LIKE 'M%' ")

# Convert the scan results into a Pandas dataframe
df = sc.to_arrow().to_pandas()
df
```

This will return only the rows where the last name starts with 'M', which includes users like *McMurrugh* and *Matschoss*.

CHAPTER 2   IMPLEMENTING THE DATA LAKEHOUSE

```
[7]: from pyiceberg.catalog import load_catalog

# Load the REST catalog
catalog = load_catalog('default')

# Load the users table
tbl = catalog.load_table('accounts.users')

# Create a scan with filter on last_name
sc = tbl.scan(row_filter="last_name LIKE 'M%' ")
# Convert the scan results into a Pandas dataframe
df = sc.to_arrow().to_pandas()
df
```

[7]:

	id	first_name	last_name	email
0	6	Calypso	McMurrugh	cmcmurrugh5@cam.ac.uk
1	8	Jack	Matschoss	jmatschoss7@latimes.com

Figure 2-7. Rows where the last name starts with 'M'

This query takes advantage of the partition we created since we are filtering by the last_name field.

Let's break down what the above code does. First, we import the pyiceberg module, followed by creating a connection to the REST catalog using load_catalog('default')

Earlier, when creating the spark-iceberg image, we copied the .pyiceberg.yaml file to the root folder of the container. That file defined the default catalog with the following configuration.

catalog:
    default:
        uri: <http://rest:8181>
        s3.endpoint: <http://minio:9000>
        s3.access-key-id: admin
        s3.secret-access-key: password

CHAPTER 2   IMPLEMENTING THE DATA LAKEHOUSE

This configuration enables PyIceberg to communicate with both the REST catalog service for metadata management and MinIO for actual data storage.

Once the catalog is loaded, we load our users table from the accounts database using `catalog.load_table('accounts.users')`. Finally, we convert the scan results into an Apache Arrow table and then into a Pandas DataFrame for easy data manipulation and analysis

Working with PyIceberg CLI

Now, let's explore another powerful tool in the Iceberg ecosystem – the PyIceberg CLI. This command-line interface provides a convenient way to inspect and manage Iceberg tables directly from your terminal. It's particularly useful for quick table inspections, schema validations, and metadata analysis without writing any code. The CLI is automatically installed when you install PyIceberg, making it readily available for administrative tasks and troubleshooting.

To list the databases available in the default catalog, run the following command in a terminal from where you launched Docker Compose. The CLI uses the `.pyiceberg.yaml` file in the root folder to locate catalog definitions.

```
> docker exec -it spark-iceberg pyiceberg list
accounts
```

To list the tables under a database:

```
> docker exec -it spark-iceberg pyiceberg list accounts
accounts.users
```

Finally, the following describes the users table.

```
> docker exec -it spark-iceberg pyiceberg describe --entity=table accounts.users
```

CHAPTER 2 IMPLEMENTING THE DATA LAKEHOUSE

```
Table format version   2
Metadata location      s3://warehouse/accounts/users/metadat
a/00001-78f6e409-71c9-4daa-a13d-16e742e5628e.metadata.json
Table UUID             a342220e-a732-47e6-9128-a5729f682247
Last Updated           1737955937100
Partition spec         [
                           1000: last_name_trunc: truncate[1](3)
                       ]
Sort order             []
Current schema         Schema, id=0
                       ├── 1: id: optional int
                       ├── 2: first_name: optional string
                       ├── 3: last_name: optional string
                       └── 4: email: optional string
Current snapshot       Operation.APPEND: id=80856117289882055,
schema_id=0
Snapshots              Snapshots
                       └── Snapshot 80856117289882055,
schema 0: s3://warehouse/accounts/users/metadata/
snap-80856117289882055-1-0f5ec793-1625-4781-
a3f9-697814a5ae4c.avro
Properties             owner                              root
                       write.parquet.compression-codec   zstd
```

You will notice a few important information in the output, including partition specification, current table schema, and the current snapshot pointer.

Summary

In this chapter, we've explored how to set up a local Iceberg lakehouse environment using Docker Compose, combining essential components like Spark, MinIO, PyIceberg, and the REST Catalog.

While this setup is perfect for local development and learning purposes, it's important to note that production deployments require more robust solutions. For production environments, we recommend a distributed deployment, preferably on a cloud platform that provides built-in scalability, security, and maintenance features.

The concepts and operations we've covered here – from table creation and data insertion to querying with PyIceberg – remain consistent whether you're working locally or in the cloud. This makes our Docker-based setup an excellent starting point for developing lakehouse solutions that can be seamlessly migrated to production environments.

In the next chapter, we will extend the lakehouse by including source systems and developing an ETL pipeline for analytics.

CHAPTER 3

Batch ETL Pipeline with Apache Spark

In the previous chapter, we explored the local development setup of the OneShop Iceberg lakehouse. Now that the data lakehouse infrastructure is fully functional, the OneShop data engineering team is planning its next steps.

The lakehouse currently stands empty. The team's first task is to "hydrate" the lakehouse by loading data into it. In this chapter, we'll create loading jobs to transfer data from various source systems into Iceberg tables using PySpark. We'll introduce PySpark for data processing, establish proper data modeling techniques for Iceberg tables, and implement loading and cleaning jobs. Before loading the data, we'll model the Iceberg tables following the Medallion architecture with bronze, silver, and gold layers to ensure data quality and accessibility.

CHAPTER 3   BATCH ETL PIPELINE WITH APACHE SPARK

Hydrating the OneShop Lakehouse

As mentioned in **Chapter 1**, OneShop primarily has two source systems to load data from:

- **Postgres database**: This is an OLTP database storing key business data in structured format, including:
    - `users` table: Information about customers who shop on the platform.
    - `items` table: A catalog of more than 1,000 products available for sale.
    - `purchases` table: Records of transactions, capturing details about what customers buy and when.
- **MinIO data lake**: The user interaction events of the store website are captured and stored in a MinIO storage bucket. The main event type currently captured is page views, formatted in JSON.

The source systems contain both structured data (Postgres relational database) and semi-structured data (JSON events in MinIO). We'll need to transform this data into columnar Parquet format and load it into Iceberg tables optimized for OLAP queries. We will write ETL (extract, transform, load) jobs in PySpark to handle these transformations.

A data lakehouse is a vast structure shared across multiple departments and user roles within an organization. Therefore, logical partitioning in the lakehouse is crucial for maintaining data quality and clear ownership. Before creating our destination Iceberg tables, we need to model the lakehouse using a standard architecture.

This is where the Medallion architecture comes in.

Medallion Architecture

The Medallion architecture, also known as the multi-hop architecture, is a data modeling pattern commonly used in data lakehouses to organize data based on its quality, reliability, and level of transformation. This architecture helps maintain data quality, enables efficient data governance, and provides a clear path for data transformation.

The architecture consists of three main layers, often represented by bronze, silver, and gold medals:

Bronze Layer (Raw)

The Bronze layer is where raw data first enters the lakehouse, preserved exactly as it comes from source systems. This layer maintains data in its unprocessed, original format, serving as a crucial backup of source data. You'll find raw JSON events from website logs and database dumps stored here.

Silver Layer (Validated)

In the Silver layer, data undergoes its initial processing and validation phase. Here, data is cleaned, filtered, and made to conform to a defined schema. This process includes removing duplicate records and standardizing data types. The result includes properly parsed JSON events with correct data types and validated customer records.

Gold Layer (Business)

The Gold layer represents the final stage, where data is fully curated and ready for business use. This layer houses aggregated and enriched data models, specifically optimized for various use cases and analysis. Common examples include Customer 360° views, daily sales aggregations, and comprehensive product analytics.

CHAPTER 3   BATCH ETL PIPELINE WITH APACHE SPARK

This layered approach offers several key advantages: it enables clear data lineage and traceability, ensures improved data quality through progressive refinement, delivers better performance for end-user queries, and facilitates simplified debugging and data recovery processes.

For OneShop's lakehouse implementation, we'll follow this architecture to ensure our data is properly organized and maintained throughout its lifecycle.

Before You Begin

You will find the code for this chapter located in the `chapter-03` folder. If you haven't set everything up yet, refer to the **Prerequisites** section in the first chapter for more information.

Navigate to the project folder on a terminal by typing:

```
cd <repository_root>/chapter-03
```

Docker Setup Overview

The `chapter-03/docker-compose.yaml` file is nearly identical to the one in `chapter-02`, with the addition of two new service definitions: `postgres` and `loadgen`.

The `postgres` service mimics the Postgres database currently used at OneShop. It is the source system of the ETL pipeline.

```
postgres:
    image: postgres:16
    hostname: postgres
    container_name: postgres
    networks:
      iceberg_net:
```

CHAPTER 3   BATCH ETL PIPELINE WITH APACHE SPARK

```
ports:
  - 5432:5432
environment:
  - POSTGRES_USER=postgresuser
  - POSTGRES_PASSWORD=postgrespw
  - POSTGRES_DB=oneshop
  - PGPASSWORD=postgrespw
volumes:
  - ./postgres/postgres_bootstrap.sql:/docker-entrypoint-
    initdb.d/postgres_bootstrap.sql
healthcheck:
  test: ["CMD-SHELL", "pg_isready -U postgresuser -d
  oneshop"]
  interval: 1s
  start_period: 60s
```

The above YAML configuration starts a Postgres container service from the postgres:16 base image. The environment section defines the standard Postgres environment variables during the initialization.

1. POSTGRES_USER and POSTGRES_PASSWORD set the username and password for Postgres, respectively.

2. POSTGRES_PASSWORD=postgrespw: This variable sets the password for the PostgreSQL user. Here, the password is set to postgrespw.

3. POSTGRES_DB=oneshop: This variable specifies the name of the database to be created when the Postgres container starts. The database name is set to oneshop.

53

4. PGPASSWORD=postgrespw: This environment variable sets the password for the PostgreSQL client authentication. It is used by client applications to connect to the PostgreSQL database. In this case, it is set to the same value as POSTGRES_PASSWORD, which is postgrespw.

In the volumes section, we mount a SQL script at ./postgres/postgres_bootstrap.sql that executes during database startup to create users, items, and purchases tables. You will find this script inside the ./chapter-03/postgres folder.

We will explore the table schemas later in the chapter. For now, notice how we created the readonly user role, granted it the necessary privileges, and assigned a new user, etluser, to that role inside ./postgres/postgres_bootstrap.sql.

```sql
CREATE USER etluser WITH PASSWORD 'etlpassword';

-- Create a readonly role
CREATE ROLE readonly;

-- Grant privileges to the readonly role
GRANT CONNECT ON DATABASE oneshop TO readonly;
GRANT USAGE ON SCHEMA public TO readonly;
GRANT SELECT ON ALL TABLES IN SCHEMA public TO readonly;

-- Ensure future tables also grant SELECT to readonly
ALTER DEFAULT PRIVILEGES IN SCHEMA public GRANT SELECT ON TABLES TO readonly;

-- Assign the readonly role to etluser
GRANT readonly TO etluser;

-- Grant SELECT permission on the newly created tables
GRANT SELECT ON TABLE users TO readonly;
```

CHAPTER 3   BATCH ETL PIPELINE WITH APACHE SPARK

```
GRANT SELECT ON TABLE items TO readonly;
GRANT SELECT ON TABLE purchases TO readonly;
```

In summary, the `readonly` role only has privileges to connect to the `oneshop` database, access its objects, and run SELECT queries on all tables. The `etluser` credentials will be used by all the PySpark ETL scripts we discuss later in this chapter.

Granting only scoped permissions to ETL pipeline accounts is a crucial security practice for several reasons:

- **Minimizes attack surface:** If the ETL account credentials are compromised, the attacker can only read data, not modify or delete it

- **Follows principle of least privilege:** The ETL process only needs read access to perform its job, so it shouldn't have unnecessary additional permissions

- **Reduces risk of accidental changes:** Even if there's a bug in the ETL code, it cannot accidentally modify or delete production data

- **Enables better auditing:** Having dedicated read-only accounts makes it easier to track and audit data access patterns

Next, we have the `loadgen` service, which populates the `oneshop` database tables with mock data.

```
loadgen:
  build: loadgen
  container_name: loadgen
  init: true
  networks:
    iceberg_net:
  depends_on:
    postgres: {condition: service_healthy}
```

The `./chapter-03/loadgen` folder contains a `Dockerfile` that provides instructions to create a Python container and install required dependencies. When this container starts, it executes the Python script `generate_load.py`, which performs two tasks:

1. Populates the `users`, `items`, and `purchases` tables with mock data to provide a working dataset.
2. Creates the `pageviews` bucket in MinIO and fills it with mock JSON events, which we'll explore later.

As the code in `generate_load.py` handles basic tasks that fall outside the scope of this chapter, we won't delve into its details now.

With that overview complete, let's proceed to start the setup.

Running the Setup

From the root level of the `chapter-03` directory, start all containers by running:

```
docker-compose up -d --build
```

This will build the `loadgen` container first and start the `postgres` container along with the Iceberg, MinIO, and Spark containers we discussed in the previous chapter. The `loadgen` container will run briefly until it completes executing `generate_load.py`.

You can verify the database table creation by logging into Postgres:

```
dockercompose exec postgres psql -U etluser oneshop
```

CHAPTER 3 BATCH ETL PIPELINE WITH APACHE SPARK

List the tables inside oneshop

```
oneshop=> \dt;
          List of relations
 Schema |   Name    | Type  |   Owner
--------+-----------+-------+--------------
 public | items     | table | postgresuser
 public | purchases | table | postgresuser
 public | users     | table | postgresuser
```

To verify the content inside the purchases table, run:

```
oneshop=> select * from purchases limit 10;
id | user_id | item_id | quantity | purchase_price |   created_
at           |         updated_at
----+---------+---------+----------+----------------+----------
------------+----------------------------
  1 |    1798 |     552 |        3 |         458.73 | 2025-06-24
12:17:36 | 2025-06-25 09:53:57.339988
  2 |    2325 |     845 |        2 |         999.72 | 2025-06-24
11:28:52 | 2025-06-25 09:53:57.524961
  3 |    6397 |     128 |        3 |          98.91 | 2025-06-24
15:19:20 | 2025-06-25 09:53:57.670513
  4 |     236 |     439 |        3 |         692.31 | 2025-06-25
08:49:35 | 2025-06-25 09:53:57.815307
  5 |    3089 |     921 |        4 |        1024.56 | 2025-06-24
23:10:51 | 2025-06-25 09:53:57.960379
  6 |    2926 |     128 |        5 |         164.85 | 2025-06-24
22:41:54 | 2025-06-25 09:53:58.10299
  7 |    5726 |     820 |        4 |        1595.64 | 2025-06-24
15:52:38 | 2025-06-25 09:53:58.243444
  8 |    5321 |     941 |        2 |          85.14 | 2025-06-24
20:59:18 | 2025-06-25 09:53:58.386083
```

CHAPTER 3   BATCH ETL PIPELINE WITH APACHE SPARK

```
    9 |    9545 |     867 |           5 |          1384.30 | 2025-06-24 20:55:16 | 2025-06-25 09:53:58.529615
   10 |     589 |     842 |           3 |          1331.97 | 2025-06-24 18:12:11 | 2025-06-25 09:53:58.672363
(10 rows)
```

You can perform the same checks for the remaining tables as well.

Next, login to MinIO console by visiting http://localhost:9001. You should see the new pageviews bucket inside MinIO populated with events like this:

Figure 3-1. pageviews bucket inside MinIO

If everything looks correct up to this point, your setup is working properly. Otherwise, check the container logs for any errors.

Just before we move to the next section, run the following "curl" command to download the AWS jar file, which will be useful later.

```
curl -L -o ./spark/jars/aws-java-sdk-bundle-1.11.1026.jar https://repo1.maven.org/maven2/com/amazonaws/aws-java-sdk-bundle/1.11.1026/aws-java-sdk-bundle-1.11.1026.jar
```

Modeling Iceberg Tables in the Lakehouse

Let's kick things off by modeling and creating Iceberg tables in the lakehouse. To keep things simple, we will only model bronze and silver tables in this chapter, while we leave the gold tables to the next chapter.

CHAPTER 3   BATCH ETL PIPELINE WITH APACHE SPARK

We've created a Jupyter notebook with all the DDL statements written in Spark SQL, located in ./notebooks/create_iceberg_tables.ipynb. This notebook contains Spark SQL code that creates both the namespaces and the bronze and silver tables in the lakehouse.

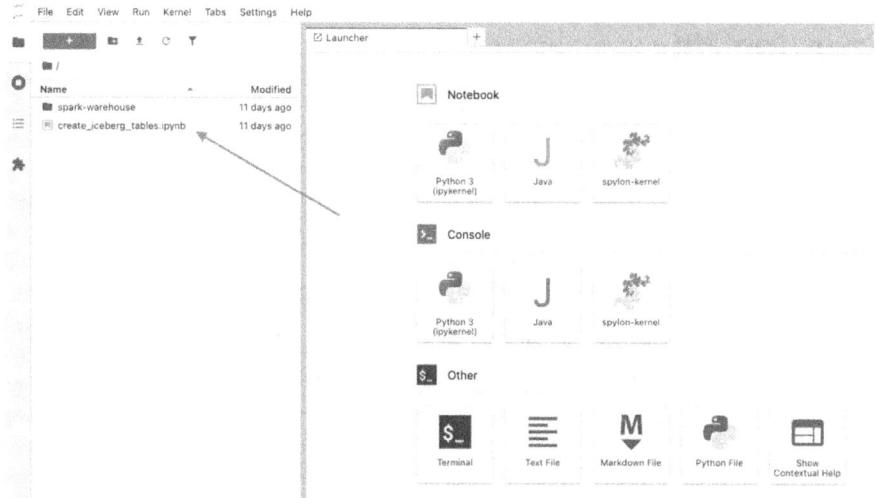

Figure 3-2. *Creating iceberg tables in the lakehouse*

To read the code:

- Open the Jupyter Lab interface at http://localhost:8888 in your browser
- Navigate to the create_iceberg_tables.ipynb notebook in the file browser

For the sake of simplicity and readability, the code in the notebook uses plain SQL rather than PySpark's DataFrame API. This is possible because we're using the %%sql magic command in Jupyter, which allows us to write standard SQL queries that are executed directly against the Spark context. This approach makes the DDL statements more concise and easier to understand, especially for those familiar with SQL but not necessarily with Spark's DataFrame API.

59

Before diving into the notebook, let's first examine the code to understand its functionality.

We start by creating three namespaces that logically partition the lakehouse into three Medallion layers.

Creating Namespaces

In Iceberg, a namespace is a logical container that organizes tables into groups, similar to how folders organize files in a file system. Namespaces help maintain a clean, hierarchical structure in your data lake, making it easier to manage access controls and organize related tables.

When working with Apache Spark, Iceberg namespaces are directly mapped to Spark databases. This means that when you create an Iceberg namespace, it appears as a database in Spark's catalog. This mapping allows for seamless integration between Spark's SQL interface and Iceberg's table management capabilities.

The following Spark SQL statements create three Iceberg namespaces that logically divide the lakehouse into three layers according to the Medallion architecture.

```
CREATE DATABASE IF NOT EXISTS bronze;
CREATE DATABASE IF NOT EXISTS silver;
CREATE DATABASE IF NOT EXISTS gold;
```

Modeling Bronze Tables

Tables in the bronze layer typically have a 1:1 schema mapping with their source database or system.

In the postgres/postgres_bootstrap.sql file you will find the Postgres schemas for users, items, and purchases source tables.

```sql
CREATE TABLE IF NOT EXISTS users
(
    id SERIAL PRIMARY KEY,
    first_name VARCHAR(100),
    last_name VARCHAR(100),
    email VARCHAR(255),
    created_at TIMESTAMP DEFAULT CURRENT_TIMESTAMP,
    updated_at TIMESTAMP DEFAULT CURRENT_TIMESTAMP
);
CREATE TABLE IF NOT EXISTS items
(
    id SERIAL PRIMARY KEY,
    name VARCHAR(100),
    category VARCHAR(100),
    price DECIMAL(7,2),
    inventory INT,
    created_at TIMESTAMP DEFAULT CURRENT_TIMESTAMP,
    updated_at TIMESTAMP DEFAULT CURRENT_TIMESTAMP
);
CREATE TABLE IF NOT EXISTS purchases
(
    id SERIAL PRIMARY KEY,
    user_id BIGINT REFERENCES users(id),
    item_id BIGINT REFERENCES items(id),
    quantity INT DEFAULT 1,
    purchase_price DECIMAL(12,2),
    created_at TIMESTAMP DEFAULT CURRENT_TIMESTAMP,
    updated_at TIMESTAMP DEFAULT CURRENT_TIMESTAMP
);
```

CHAPTER 3   BATCH ETL PIPELINE WITH APACHE SPARK

Based on the above schema, we can start modeling equivalent dimensional Iceberg tables in the bronze layer. We are using Spark SQL here due to its declarative nature and direct integration with Iceberg.

```
CREATE TABLE IF NOT EXISTS bronze.users (
    id BIGINT,
    first_name STRING,
    last_name STRING,
    email STRING,
    created_at TIMESTAMP,
    updated_at TIMESTAMP
)
USING iceberg
PARTITIONED BY (days(created_at))
TBLPROPERTIES (
    'format-version' = '2',
    'comment' = 'Dimension table for user information'
);

CREATE TABLE IF NOT EXISTS bronze.items (
    id BIGINT,
    name STRING,
    category STRING,
    price DECIMAL(7,2),
    inventory INT,
    created_at TIMESTAMP,
    updated_at TIMESTAMP
)
USING iceberg
PARTITIONED BY (category)
TBLPROPERTIES (
    'format-version' = '2',
    'comment' = 'Dimension table for product items'
);
```

Notice how users and items tables have a direct 1:1 mapping with their corresponding source tables in Postgres. We've implemented strategic partitioning to optimize query performance: the users table is partitioned by creation time (using the days function on created_at), while the items table is partitioned by category. This partitioning strategy allows for more efficient queries when filtering by these commonly used fields.

We've also specified "format-version" = "2" in the table properties, which enables Iceberg's advanced features like row-level deletes and schema evolution capabilities. This future-proofs our tables for when we need to implement data correction or schema changes.

Next, we can model the fact tables, purchases, and pageviews as follows.

```
CREATE TABLE IF NOT EXISTS bronze.purchases (
    id BIGINT,
    user_id BIGINT,
    item_id BIGINT,
    quantity INT,
    purchase_price DECIMAL(12,2),
    created_at TIMESTAMP,
    updated_at TIMESTAMP
)
USING iceberg
PARTITIONED BY (days(created_at))
TBLPROPERTIES (
    'format-version' = '2',
    'comment' = 'Fact table for purchase transactions'
);

CREATE TABLE IF NOT EXISTS bronze.pageviews (
    user_id BIGINT,
```

```
    url STRING,
    channel STRING,
    received_at TIMESTAMP
)
USING iceberg
PARTITIONED BY (days(received_at))
TBLPROPERTIES (
    'format-version' = '2',
    'comment' = 'Fact table for purchase transactions'
);
```

While the `purchases` table is loaded from Postgres, the `pageviews` table is supposed to be populated from the JSON events in MinIO.

The format of a sample pageview event looks like this:

```
{"user_id": 1, "url": "/products/1", "channel": "web", "received_at": "2021-10-01T00:00:00Z"}
```

As you can see, we modeled the `bronze.pageviews` table to closely resemble this structure. This approach allows us to ingest the raw event data with minimal transformation, preserving the original format while still gaining the benefits of Iceberg's table format, like partitioning and metadata management.

Modeling Silver Tables

Now that we've defined our bronze tables to capture the raw data, let's create the corresponding silver tables that will hold the validated and enriched data.

```
CREATE TABLE IF NOT EXISTS silver.users (
    id BIGINT,
    first_name STRING,
    last_name STRING,
```

```
    email STRING,
    created_at TIMESTAMP,
    updated_at TIMESTAMP,
    valid_email BOOLEAN,
    full_name STRING
)
USING iceberg
PARTITIONED BY (days(created_at))
TBLPROPERTIES (
    'format-version' = '2',
    'comment' = 'Validated dimension table for user information'
);

CREATE TABLE IF NOT EXISTS silver.items (
    id BIGINT,
    name STRING,
    category STRING,
    price DECIMAL(7,2),
    inventory INT,
    created_at TIMESTAMP,
    updated_at TIMESTAMP
)
USING iceberg
PARTITIONED BY (category)
TBLPROPERTIES (
    'format-version' = '2',
    'comment' = 'Validated dimension table for product items'
);
```

Notice how the silver tables enhance the bronze tables with additional derived columns. For example, we've added a full_name field to the users table that concatenates first and last names and a valid_email field to identify null and malformed email addresses.

CHAPTER 3 BATCH ETL PIPELINE WITH APACHE SPARK

```
CREATE TABLE IF NOT EXISTS silver.purchases_enriched (
    id BIGINT,
    user_id BIGINT,
    item_id BIGINT,
    quantity INT,
    purchase_price DECIMAL(12,2),
    total_price DECIMAL(14,2),
    user_email STRING,
    item_name STRING,
    item_category STRING,
    purchase_date DATE,
    purchase_hour INT,
    created_at TIMESTAMP,
    updated_at TIMESTAMP
)
USING iceberg
PARTITIONED BY (days(created_at))
TBLPROPERTIES (
    'format-version' = '2',
    'comment' = 'Validated and enriched fact table for purchase
    transactions'
);

CREATE TABLE IF NOT EXISTS silver.pageviews_by_items (
    user_id BIGINT,
    item_id BIGINT,
    page STRING,
    item_name STRING,
    item_category STRING,
    channel STRING,
    received_at TIMESTAMP
)
```

```
USING iceberg
PARTITIONED BY (days(received_at))
TBLPROPERTIES (
    'format-version' = '2',
    'comment' = 'Fact table for purchase transactions'
);
```

In the fact tables, we've added even more enrichments. The purchases_enriched table includes total_price (calculated from quantity × price), along with denormalized fields from dimension tables (user_email, item_name, and item_category). We've also added purchase_date and purchase_hour derived from created_at to make time-based querying easier. Similarly, the pageviews_by_items table extracts item_id from URLs and joins with dimension data to add context.

These enrichments serve several important purposes:

- **Performance optimization:** Pre-calculating common metrics and denormalizing dimension data reduces the need for expensive joins during analysis

- **Simplified analytics:** Business users can work with ready-to-use fields without needing complex transformations

- **Data quality improvement:** The silver layer validates and standardizes data, ensuring consistency

- **Business context:** Adding descriptive fields makes the data more meaningful and accessible to non-technical users

This pattern of progressive refinement is core to the Medallion architecture, where each layer adds more business value to the raw data.

CHAPTER 3 BATCH ETL PIPELINE WITH APACHE SPARK

Creating Iceberg Tables by Running the Notebook

Now that we've examined the code for creating our Iceberg tables, let's proceed with actually creating these tables in the lakehouse. We'll use the Jupyter notebook we discussed, which contains all the DDL statements we've reviewed.

To create the tables:

- Open the Jupyter Lab interface at http://localhost:8888 in your browser

- Navigate to the create_iceberg_tables.ipynb notebook in the file browser

- Run all cells in the notebook by selecting "Run" ➤ "Run All Cells" from the menu

This will execute all the CREATE TABLE statements through PySpark and establish our bronze and silver layer tables in the lakehouse structure we've designed.

CHAPTER 3   BATCH ETL PIPELINE WITH APACHE SPARK

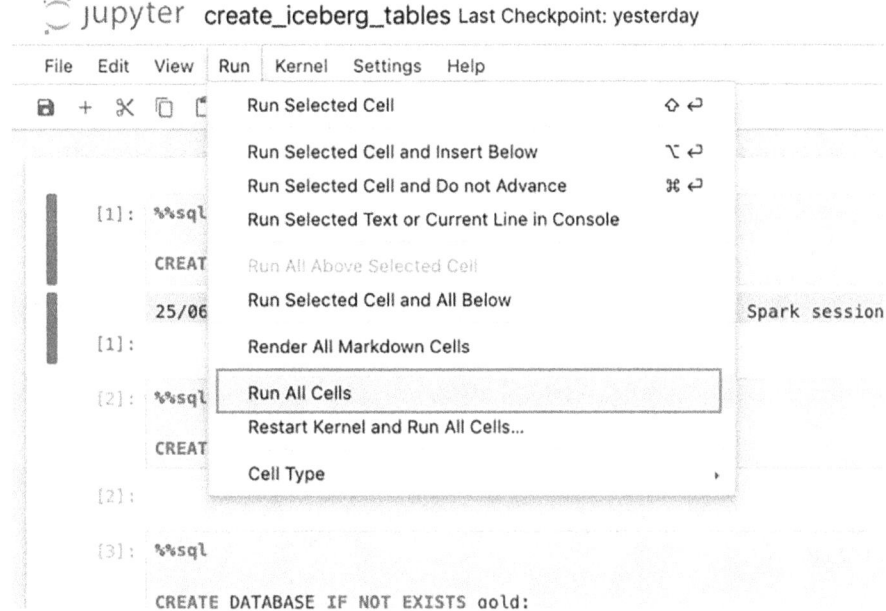

Figure 3-3. Running the notebook

Validating Table Creation

After running the notebook, you should validate that all tables were created successfully. You can do this by executing the following SQL queries in a new notebook cell.

Note that you'll need to prepend the magic command "%%sql" to each cell when using plain SQL like this.

CHAPTER 3 BATCH ETL PIPELINE WITH APACHE SPARK

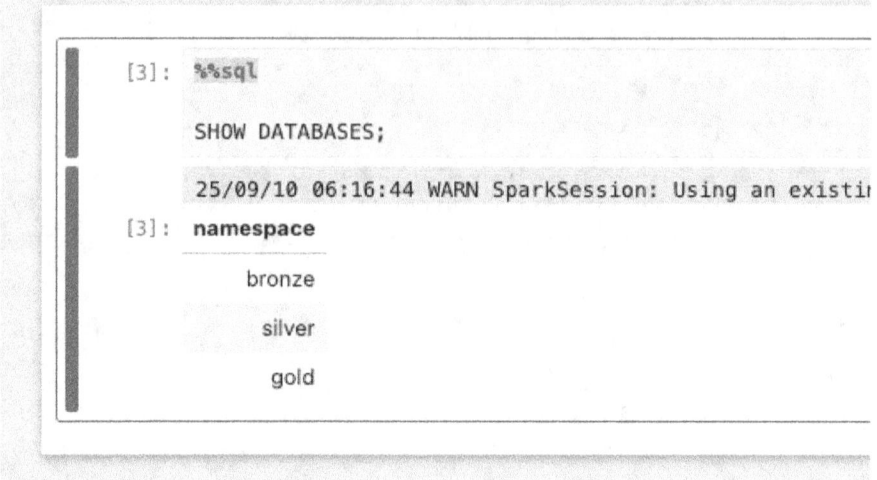

Figure 3-4. *Executing the SQL queries directly in notebook cells with JupySQL*

```
-- List all databases (namespaces)
SHOW DATABASES;

-- List all tables in the bronze namespace
SHOW TABLES IN bronze;

-- List all tables in the silver namespace
SHOW TABLES IN silver;

-- Examine the schema of a specific table
DESCRIBE TABLE bronze.users;
```

You should see output confirming the creation of all three namespaces (bronze, silver, gold) and all the tables we defined. The DESCRIBE command will show you the detailed schema of any table, including column names, data types, and comments.

CHAPTER 3　BATCH ETL PIPELINE WITH APACHE SPARK

Additionally, you can run a simple query to verify that the tables are properly configured:

```
-- Check the table properties, including partitioning
SELECT * FROM bronze.users.properties;

-- Similar checks can be performed for other tables
SELECT * FROM iceberg.silver.purchases_enriched.properties;
```

This validation step ensures that all tables are created with the correct schema, partitioning strategy, and table properties before proceeding to the data loading phase.

Loading Iceberg Tables from Postgres and MinIO

With our table structure defined for both bronze and silver layers, we're now ready to write the PySpark jobs that will extract data from our source systems and load it into these Iceberg tables

Loading Postgres Tables

The PySpark script `spark/scripts/postgres_loader.py` handles the extract, transform, and load (ETL) process that transfers data from Postgres to the data lakehouse. It extracts the `users`, `items`, and `purchases` tables from Postgres, converts them to Iceberg format, and loads them into their corresponding Iceberg tables.

Let's break down the script and understand it.

```python
import sys
from pyspark.sql import SparkSession
from pyspark.sql.functions import col
```

CHAPTER 3   BATCH ETL PIPELINE WITH APACHE SPARK

```
POSTGRES_URL = "jdbc:postgresql://postgres:5432/oneshop"
USERNAME = "etluser"
PASSWORD = "etlpassword"

try:
    spark = SparkSession.builder \\
        .appName("postgres-to-iceberg-loader") \\
        .getOrCreate()
except Exception as e:
    print(f"Error creating SparkSession: {e}")
    sys.exit(1)

print("SparkSession created successfully.")
```

This code initializes a PySpark application by creating a SparkSession, which is the entry point for any Spark functionality. It includes error handling to catch any initialization issues. The defined parameters set up configuration variables for the source PostgreSQL connection (URL, username, and password).

Notice how the application is using the dedicated `etluser` credentials to connect to the Postgres database. This dedicated user account we provisioned earlier in this chapter for ETL processes provides better security through the principle of least privilege, allowing you to grant only the specific permissions needed for data extraction.

This is just the setup portion of the script. The next sections contain the actual data extraction from Postgres and loading into Iceberg tables.

```
# 1. Load users table
print("Processing 'users' table...")
try:
    users_df = spark.read \\
        .format("jdbc") \\
        .option("driver", "org.postgresql.Driver") \\
        .option("url", POSTGRES_URL) \\
```

```
        .option("dbtable", "users") \\
        .option("user", USERNAME) \\
        .option("password", PASSWORD) \\
        .load()

    # Ensure column types match Iceberg schema (optional, but
      good practice)
    # Cast necessary columns if there are discrepancies, e.g.,
      if Postgres SERIAL was read as INT
    users_df = users_df.select(
        col("id").cast("long"),
        col("first_name").cast("string"),
        col("last_name").cast("string"),
        col("email").cast("string"),
        col("created_at").cast("timestamp"),
        col("updated_at").cast("timestamp")
    )

    # Write to Iceberg table. 'overwrite' mode is used for
      initial load.
    users_df.write \\
        .format("iceberg") \\
        .mode("overwrite") \\
        .option("format-version", "2") \\
        .save("analytics.dim_users")
    print("'users' table loaded successfully into Iceberg.")
except Exception as e:
    print(f"Error loading 'users' table: {e}")
```

The above code loads the users table into bronze.users Iceberg table. Let's break it down further.

First, PySpark uses the Spark JDBC connector to read from the 'users' table in PostgreSQL. It then explicitly casts each column to ensure type compatibility between Postgres and Iceberg schemas. This is a best practice to prevent data type mismatches during the transfer process.

The key part is the `users_df.write()` function where the DataFrame is written into the destination `bronze.users` table. The `format("iceberg")` option ensures that the data is stored in the Apache Iceberg table format while replace any existing data in it.

The rest of the script continues with loading the `items` and `purchases` tables, following a similar pattern. Each section reads from the corresponding Postgres table, ensures data type compatibility, and writes to its Iceberg destination.

Finally, let's execute this script by running the following:

```
docker compose exec spark-iceberg /opt/spark/bin/spark-submit \\
--jars /home/iceberg/pyspark/jars/postgresql-42.7.6.jar \\
/home/iceberg/pyspark/scripts/postgres_loader.py
```

Notice we have included the Postgres JDBC connector in the --jars option. We copied this jar file into the spark-iceberg container as a volume mount. Otherwise, this code won't run.

Loading PageView Events from MinIO

After loading raw Postgres tables into the bronze layer, the next step is to load the JSON-formatted PageView events in the MinIO bucket into the `brone.pageviews` table.

Let's look at the PySpark script that handles this loading job: `spark/scripts/minio_loader.py`.

The first section of the script initializes a Spark session and establishes a connection to the MinIO storage bucket named pageviews using the provided configuration parameters. The minio_endpoint variable specifies the hostname of the MinIO container that runs on port 9000.

```
import sys
from pyspark.sql import SparkSession
from pyspark.sql.functions import col

minio_access_key = "admin"
minio_secret_key = "password"
minio_endpoint = "<http://minio:9000>"
minio_bucket = "pageviews"

try:
    spark = SparkSession.builder \\
        .appName("minio-to-iceberg-loader") \\
        .config("spark.hadoop.fs.s3a.endpoint", minio_
        endpoint) \\
        .config("spark.hadoop.fs.s3a.access.key", minio_
        access_key) \\
        .config("spark.hadoop.fs.s3a.secret.key", minio_
        secret_key) \\
        .config("spark.hadoop.fs.s3a.path.style.access",
        "true") \\
        .config("spark.hadoop.fs.s3a.impl", "org.apache.hadoop.
        fs.s3a.S3AFileSystem") \\
        .getOrCreate()
except Exception as e:
    print(f"Error creating SparkSession: {e}")
    sys.exit(1)

print("SparkSession created successfully.")
```

This Spark job uses the **S3AFileSystem** plugin to read from the S3-compatible MinIO storage. This is configured in the SparkSession builder with:

```
.config("spark.hadoop.fs.s3a.impl", "org.apache.hadoop.fs.s3a.S3AFileSystem")
```

This plugin allows Spark to interact with S3-compatible object storage systems like MinIO using the s3a:// protocol, with additional configurations for authentication and endpoint specification.

The next section shows us how Spark can just as easily read from JSON files in the storage bucket as it can from relational databases. It reads the JSON files, applies the necessary type conversions, and loads the data into the bronze.pageviews Iceberg table.

```
try:
    print("Reading pageview events from MinIO...")
    pageviews_df = spark.read.json(f"s3a://{minio_bucket}/")

    pageviews_df = pageviews_df.select(
        col("user_id").cast("long"),
        col("url").cast("string"),
        col("channel").cast("string"),
        col("received_at").cast("timestamp")
    )

    # Write to Iceberg table. 'overwrite' mode is used for
      initial load.
    pageviews_df.write \\
        .format("iceberg") \\
        .mode("overwrite") \\
        .option("format-version", "2") \\
        .save("bronze.pageviews")
```

```
    print("'pageviews' bucket content loaded successfully into
    Iceberg.")
except Exception as e:
    print(f"Error processing pageview events: {e}")
    sys.exit(1)
```

Finally, let's execute this script by running the following:

```
docker compose exec spark-iceberg /opt/spark/bin/spark-submit \\
--jars /home/iceberg/pyspark/jars/hadoop-aws-3.3.4.jar,/home/iceberg/pyspark/jars/aws-java-sdk-bundle-1.11.1026.jar \\
/home/iceberg/pyspark/scripts/minio_loader.py
```

Notice how we included the dependent plugins under --jars options. The Hadoop AWS and AWS Java SDK Bundle JARs are required for the S3A connector to function properly when connecting to S3-compatible storage like MinIO.

Note that separating the loading jobs into two distinct scripts offers several benefits. It allows different teams to take ownership of specific data domains, enables independent scaling based on source system requirements, simplifies maintenance, and provides better error isolation. While this separation is optional, it follows the best practice of modular ETL pipeline design.

Data Cleaning, Denormalization, and Enrichment with PySpark

After loading our raw data into the bronze layer, the next step is to clean and transform it before moving to the silver layer.

CHAPTER 3 BATCH ETL PIPELINE WITH APACHE SPARK

Let's examine key portions of the PySpark script that handle this transformation:

```
spark/scripts/bronze_to_silver_transformer.py.
from pyspark.sql import SparkSession
from pyspark.sql.functions import col, concat_ws, upper, regexp_extract, lit, when
from pyspark.sql.functions import hour, to_date

# Initialize Spark session with Iceberg support
spark = (
    SparkSession.builder
    .appName("bronze-to-silver-transformer")
    .config("spark.sql.catalog.spark_catalog", "org.apache.iceberg.spark.SparkSessionCatalog")
    .config("spark.sql.catalog.spark_catalog.type", "hive")
    .getOrCreate()
)

bronze_users = spark.table("bronze.users")
bronze_items = spark.table("bronze.items")
bronze_purchases = spark.table("bronze.purchases")
bronze_pageviews = spark.table("bronze.pageviews")
```

The PySpark script begins by creating a SparkSession with Iceberg support. This session serves as the entry point for all Spark operations. It then loads the four bronze layer tables – users, items, purchases, and pageviews – into DataFrames using the spark.table() method, which directly reads from the Iceberg tables.

The first transformation populates two derived columns in the users table by

- Validating user email addresses by checking them against a regular expression pattern to ensure they follow standard email format.
- Populating the full_name field by concatenating the first and last names with a space between them.

```
# Define a simple email validation regex (basic, can be
improved)
email_regex = r"^[A-Za-z0-9._%+-]+@[A-Za-z0-9.-]+\\.[A-Za-z]{2,}$"

# Transformations:
silver_users = (
    bronze_users
    .withColumn("valid_email", col("email").rlike(email_regex))
    .withColumn("full_name", concat_ws(" ", col("first_name"),
    col("last_name")))
)
```

Second, the price field is validated to ensure positive values, while the category field is transformed to uppercase.

```
silver_items = (
    bronze_items
    .withColumn("price",
        when(col("price") < 0, lit(0)).otherwise(col("price"))
    )
    .withColumn("category", upper(col("category")))
)
```

Next, several dimensional tables in the bronze layer are joined together to produce two denormalized and enriched silver tables. The silver_purchases table joins purchase transactions with user and item data to create a comprehensive view that includes

CHAPTER 3 BATCH ETL PIPELINE WITH APACHE SPARK

- Transaction details (ID, quantity, purchase price)
- Calculated total price (quantity × purchase price)
- User information (user ID and email)
- Item details (item name and category)
- Transaction timestamps (created_at), plus derived columns for purchase_date and purchase_hour for time-based analysis

Similarly, the `silver_pageviews_by_items` table processes website pageview events, extracting page name and item ID from URL paths using regular expressions, then joins with the items table to include

- User ID for tracking user behavior
- Item details (ID, name, and category) that the user viewed
- Page information indicating which section of the site was visited
- Channel information showing how users reached the page
- Timestamp of when the pageview event occurred

```
# Join and enrich the purchases table
silver_purchases = (
    bronze_purchases
    .join(bronze_users, bronze_purchases.user_id == bronze_
    users.id, "left")
    .join(bronze_items, bronze_purchases.item_id == bronze_
    items.id, "left")
    .select(
        bronze_purchases.id,
```

```
            bronze_purchases.user_id,
            bronze_purchases.item_id,
            bronze_purchases.quantity,
            bronze_purchases.purchase_price,
            (col("quantity") * col("purchase_price")).alias("total_
            price"),
            bronze_users.email.alias("user_email"),
            bronze_items.name.alias("item_name"),
            bronze_items.category.alias("item_category"),
            to_date(bronze_purchases.created_at).
            alias("purchase_date"),
            hour(bronze_purchases.created_at).
            alias("purchase_hour"),
            bronze_purchases.created_at,
            bronze_purchases.updated_at
    )
)

# The url format is "/{page_name}/{item_id}"
# Extract page_name and item_id using regex
pageviews_with_item = bronze_pageviews.withColumn(
    "page", regexp_extract(col("url"), r"^/([^/]+)/\\d+$", 1)
).withColumn(
    "item_id", regexp_extract(col("url"), r"/(\\d+)$",
1).cast("bigint")
).filter(col("item_id").isNotNull())

# Join with items to get item_name and item_category
silver_pageviews_by_items = (
    pageviews_with_item
    .join(bronze_items, pageviews_with_item.item_id == bronze_
    items.id, "left")
```

CHAPTER 3   BATCH ETL PIPELINE WITH APACHE SPARK

```
    .select(
        pageviews_with_item.user_id,
        pageviews_with_item.item_id,
        pageviews_with_item.page,
        bronze_items.name.alias("item_name"),
        bronze_items.category.alias("item_category"),
        pageviews_with_item.channel,
        pageviews_with_item.received_at
    )
)
```

Finally, the cleaned and transformed DataFrames are written to their respective silver layer tables.

For example, this is how it is done for the `silver.users` table:

```
# Write to silver.users Iceberg table (overwrite or append as needed)
(
    silver_users
    .select(
        "id",
        "first_name",
        "last_name",
        "email",
        "created_at",
        "updated_at",
        "valid_email",
        "full_name"
    )
    .writeTo("silver.users")
    .overwritePartitions()
)
```

It uses the Iceberg table format's overwritePartitions() method to update the data while maintaining table history and metadata.

In summary, the four silver layer tables being written are

- `silver.users` – Contains user data with additional derived columns for email validation and full name

- `silver.items` – Contains item data with price validation and standardized categories

- `silver.purchases_enriched` – Contains enriched purchase data joined with user and item information

- `silver.pageviews_by_items` – Contains page view data with extracted item and page information

The overwritePartitions() method is particularly useful as it allows for incremental updates by overwriting only the partitions that have changed, rather than the entire table. This is more efficient than a full overwrite, especially for large datasets where only a portion of the data changes with each update cycle.

Finally, run the script by

```
docker-compose exec spark-iceberg bash -c "cd /home/iceberg && /opt/spark/bin/spark-submit --jars
 /home/iceberg/pyspark/jars/hadoop-aws-3.3.4.jar,/home/iceberg/pyspark/jars/aws-java-sdk-bundle-1.11.1026.jar,/home/iceberg/pyspark/jars/postgresql-42.7.6.jar
 /home/iceberg/pyspark/scripts/bronze_to_silver_transformer.py"
```

With these transformations complete, the data is now in a cleaned, validated, and enriched state in the silver layer, ready for analytical queries and further transformations to the gold layer if needed.

Verify Silver Tables

Let's verify the data we've transformed and loaded into the silver layer. There are two convenient ways to inspect our Iceberg tables.

Option 1: Using a Spark Notebook

You can create a new notebook and run SQL queries to verify the data:

```
-- View all available Iceberg catalogs and namespaces
SHOW DATABASES;

-- List all tables in the silver namespace
SHOW TABLES IN silver;

-- Sample data from silver.users table
SELECT * FROM silver.users LIMIT 10;

-- Count records in silver.purchases_enriched
SELECT COUNT(*) FROM silver.purchases_enriched;

-- Check the schema of silver.pageviews_by_items
DESCRIBE TABLE silver.pageviews_by_items;
```

This allows you to quickly inspect the table structure, count records, and view sample data to ensure your transformations worked as expected.

Option 2: Using PyIceberg CLI

For a command-line approach, you can use the PyIceberg CLI to examine the Iceberg metadata and structure:

```
# List all namespaces
docker compose exec spark-iceberg pyiceberg list

# List tables in the silver namespace
docker compose exec spark-iceberg pyiceberg list silver
```

```
# Describe a specific table
docker compose exec spark-iceberg pyiceberg describe silver.purchases_enriched
```

The PyIceberg CLI provides deeper insights into Iceberg-specific details, such as table format versions, partitioning information, and metadata like schema evolution history.

Confirming Successful Data Transformation

When examining the silver tables, verify that:

- The data counts match expectations (e.g., all valid records were transferred from bronze)
- Derived columns like `valid_email` and `full_name` in `silver.users` are properly populated
- Joins produced the expected denormalized data structure in `silver.purchases_enriched`
- The `silver.pageviews_by_items` table successfully extracted and populated `page` and `item_id` from URLs

With these verification steps complete, you can be confident that your lakehouse's silver layer is ready for analytics use cases in the next chapter.

Summary

In this chapter, we tackled a critical phase of building our lakehouse architecture using Apache Iceberg. Our goal was to implement a functional data pipeline following the medallion architecture pattern, moving from raw data ingestion to transformed, analysis-ready datasets.

CHAPTER 3   BATCH ETL PIPELINE WITH APACHE SPARK

We began by modeling our Iceberg tables according to the medallion architecture's bronze and silver layers. For the bronze layer, we created tables that would store raw data in its original form but in a more query-efficient format. For the silver layer, we designed tables that would contain cleaned, validated, and enriched versions of the bronze data.

With our table structure in place, we developed ETL scripts using PySpark to extract data from operational systems (PostgreSQL database and MinIO object storage) and load it into our bronze layer. These scripts handled the initial data ingestion process, preserving the raw data while converting it to the Iceberg format.

Once the data landed in the bronze layer, we implemented additional PySpark transformations to move it to the silver layer. These transformations included:

- Data validation (email format checking, price validation)
- Data enrichment (creating full names, standardizing categories)
- Denormalization (joining multiple tables to create comprehensive views)
- Field extraction (parsing URLs to extract page names and item IDs)

We completed the process by verifying our silver layer tables using both Spark SQL queries and the PyIceberg CLI, confirming that our transformations worked as expected and the data was ready for analytical use.

In the next chapter, we'll take our lakehouse architecture to the next level by creating analytical tables in the gold layer using Trino query engine. These gold tables will aggregate and model data specifically for business intelligence needs. We'll also explore how to create visualizations from our lakehouse data, demonstrating the full value of our architecture from raw data ingestion to actionable insights.

CHAPTER 4

Data Visualization with Apache Superset

With validated, enriched, and denormalized data in the silver layer, the OneShop data engineering team plans the next steps in their lakehouse architecture – to create analytical tables in the gold layer that aggregate and model data specifically for business intelligence needs. Furthermore, they will build a business intelligence dashboard backed by these gold tables.

In this chapter, we will use Trino, an Iceberg-compatible query engine, to create gold layer tables that deliver five business metrics. Next, we will create an Apache Superset dashboard that visualizes those metrics.

KPIs for OneShop

The OneShop management expects five key business metrics in the new dashboard. We will create corresponding aggregated tables in the gold layer that delivers these metrics.

1. **Top 10 items by total revenue** – Ranks items based on total sales amount

2. **Sales performance in the last 24 hours** – Item sales revenue in the past 24 hours with hour granularity.

3. **Top converting items** – Join page views and purchases data to calculate conversion rates by item

4. **Sales volume by item category** – Item sales volume grouped by item category

5. **Page views by channel** – Show traffic sources and their contribution

Before You Begin

You will find the code for this chapter located in the chapter-04 folder. If you haven't set everything up yet, refer to the **Prerequisites** section in the first chapter for more information.

Navigate to the project folder on a terminal by typing:

```
cd <repository_root>/chapter-04
```

Docker Setup Overview

The chapter-04/docker-compose.yml file is nearly identical to the one in chapter-03, with the addition of a few new service definitions: trino, superset, and superset-db

```
trino:
    image: 'trinodb/trino'
    hostname: trino
    container_name: trino
    volumes:
        - ./trino/etc/catalog:/etc/trino/catalog
    ports:
        - '9090:8080'
```

```yaml
  networks:
    iceberg_net:

superset:
  build: ./superset
  container_name: superset
  environment:
    - SUPERSET_SECRET_KEY=mysecretkey
    - ADMIN_USERNAME=admin
    - ADMIN_PASSWORD=admin
    - ADMIN_FIRST_NAME=Superset
    - ADMIN_LAST_NAME=Admin
    - ADMIN_EMAIL=admin@example.com
    - DATABASE_URL=postgresql+psycopg2://superset:superset
      @superset-db:5432/superset
  depends_on:
    - superset-db
  ports:
    - "8088:8088"
  volumes:
    - ./superset/superset_home:/app/superset_home
  networks:
    iceberg_net:
  command: >
    /bin/sh -c "
    superset db upgrade &&
    superset fab create-admin --username admin --firstname
    Superset --lastname Admin --email admin@example.com
    --password admin || true &&
    superset init &&
    superset run -h 0.0.0.0 -p 8088
    "
```

```yaml
  superset-db:
    image: postgres:15
    container_name: superset-db
    environment:
      POSTGRES_DB: superset
      POSTGRES_USER: superset
      POSTGRES_PASSWORD: superset
    volumes:
      - ./superset/superset_db_data:/var/lib/postgresql/data
    networks:
      iceberg_net:
```

The additional three services are:

- **trino**: A distributed SQL query engine that allows us to query data from multiple sources, including our Iceberg tables.

- **superset**: An open-source business intelligence web application that enables us to create interactive dashboards and visualizations. It connects to our Trino service to query data and present it visually.

- **superset-db**: A Postgres database that stores Superset's metadata, including saved queries, dashboard configurations, and user information. This database is essential for Superset's persistence layer. Note that this is a separate database from the source Postgres.

These services complement the existing components from Chapter 3, creating a complete modern data stack with data ingestion, transformation, querying, and visualization capabilities.

CHAPTER 4 DATA VISUALIZATION WITH APACHE SUPERSET

Running Everything

From the root level of the chapter-04 directory, start all containers by running:

docker compose up -d --build

This command first builds the superset and loadgen containers and launches the entire Docker stack that we explored in **Chapter 03** with preloaded Postgres and MinIO source systems, along with Trino and Superset containers. However, we need to prepare the lakehouse same as we did in **Chapter 03** to ensure we have data available for Trino and Superset.

The ./lakehouse-preparer.sh bash script contains all the necessary commands to prepare the lakehouse before we start working on the gold layer tables.

```
#!/bin/bash

set -e

if [ ! -f ./spark/jars/aws-java-sdk-bundle-1.11.1026.jar ]; then
  echo "Downloading aws-java-sdk-bundle jar..."
  curl -L -o ./spark/jars/aws-java-sdk-bundle-1.11.1026.jar https://repo1.maven.org/maven2/com/amazonaws/aws-java-sdk-bundle/1.11.1026/aws-java-sdk-bundle-1.11.1026.jar
else
  echo "aws-java-sdk-bundle jar already exists. Skipping download."
fi

echo "Running notebook to create Iceberg tables..."
docker compose exec spark-iceberg jupyter execute /home/iceberg/notebooks/create_iceberg_tables.ipynb
```

CHAPTER 4   DATA VISUALIZATION WITH APACHE SUPERSET

```
echo "Loading data from Postgres to Iceberg..."
docker compose exec spark-iceberg /opt/spark/bin/spark-submit \\
  --jars /home/iceberg/pyspark/jars/postgresql-42.7.6.jar \\
  /home/iceberg/pyspark/scripts/postgres_loader.py

echo "Loading data from MinIO to Iceberg..."
docker compose exec spark-iceberg /opt/spark/bin/spark-submit \\
  --jars /home/iceberg/pyspark/jars/hadoop-aws-3.3.4.jar,/home/iceberg/pyspark/jars/aws-java-sdk-bundle-1.11.1026.jar \\
  /home/iceberg/pyspark/scripts/minio_loader.py

echo "Transforming bronze to silver tables..."
docker compose exec spark-iceberg /opt/spark/bin/spark-submit \\
  /home/iceberg/pyspark/scripts/bronze_to_silver_transformer.py

echo "Lakehouse preparation pipeline completed."
```

In summary, this script performs several operations inside the `spark-iceberg` container:

1. Executes the `create_iceberg_tables.ipynb` notebook via Jupyter CLI to create the Iceberg namespaces and tables in the bronze and silver layers.

2. Runs the two PySpark scripts that load data from Postgres and MinIO into the lakehouse.

3. Executes the `bronze_to_silver_transformer.py` script that validates and transforms bronze layer tables into silver tables.

Navigate to the root level of chapter-04 and execute this to kick off the lakehouse preparation procedures:

./lakehouse-preparer.sh

After this, our lakehouse will be fully hydrated with data from the bronze through the silver layer, ready for us to build our gold layer tables with Trino and create dashboards in Superset.

Create Gold Tables with Trino

Gold tables represent the final, business-ready layer in a data lakehouse architecture, specifically designed for analytical and reporting purposes. These tables contain aggregated, denormalized data that has been transformed to answer specific business questions or support particular analytics use cases. Gold tables are optimized for query performance and usability, making them ideal for business intelligence tools, dashboards, and self-service analytics.

While we could create gold tables with PySpark, data analysts who typically work in the gold layer often prefer using SQL's declarative approach rather than writing code. Therefore, we'll use this opportunity to introduce you to Trino, a SQL-compliant query engine.

Trino is a distributed SQL query engine designed for fast analytics across multiple data sources. It provides high-performance querying capabilities through a familiar SQL interface, making it ideal for data analysts working with data lakes. Trino offers native support for Apache Iceberg, allowing users to query, create, and manage Iceberg tables directly with SQL commands, which simplifies working with data lake implementations.

Trino provides a command line interface (CLI) that enables you to perform DDL operations and execute SQL queries.

CHAPTER 4   DATA VISUALIZATION WITH APACHE SUPERSET

Access the CLI by running:

```
docker compose exec trino trino
```

The SHOW CATALOGS query returns all the catalogs maintained in Trino.

```
trino> SHOW CATALOGS;
 Catalog
---------
 iceberg
 system
(2 rows)

Query 20250627_133313_00022_zf66h, FINISHED, 1 node
Splits: 19 total, 19 done (100.00%)
0.15 [0 rows, 0B] [0 rows/s, 0B/s]
```

You will notice iceberg in the list since it has already been configured. To make things simpler, we've already configured Trino to use the existing Iceberg REST catalog that we used in Chapters 2 and 3. The relevant configurations can be found in the ./trino/etc/iceberg.properties file.

```
# metastore
# <https://trino.io/docs/current/connector/iceberg.html>
connector.name=iceberg
iceberg.catalog.type=rest
iceberg.rest-catalog.uri=http://rest:8181
iceberg.file-format=parquet

# object store
# <https://trino.io/docs/current/connector/object-storage.html>
fs.hadoop.enabled=false
fs.native-s3.enabled=true
s3.endpoint=http://minio:9000
s3.region=us-east-1
```

```
s3.aws-access-key=admin
s3.aws-secret-key=password
s3.path-style-access=true
```

This configuration sets up Trino to work with Iceberg tables through a REST catalog. The key parameters include specifying `iceberg` as the connector name and setting the catalog type to `rest` with the URI pointing to the REST service. For storage, the configuration disables Hadoop filesystem support in favor of native S3 connectivity, connecting to MinIO (which provides S3-compatible storage) using the specified endpoint, access credentials, and enabling path-style access, which is required for MinIO. The configuration also sets Parquet as the default file format for Iceberg tables, which provides efficient storage and good query performance.

With Trino configured with the Iceberg catalog, we can see the existing Iceberg namespace and tables we've defined in the previous chapter.

```
# To view bronze tables
trino> use iceberg.bronze;
USE
trino:bronze> show tables;
   Table
-----------
 items
 pageviews
 purchases
 users
(4 rows)

# To view silver tables
trino:bronze> use iceberg.silver;
USE
trino:silver> show tables;
        Table
```

CHAPTER 4   DATA VISUALIZATION WITH APACHE SUPERSET

```
--------------------
 items
 pageviews_by_items
 purchases_enriched
 users
(4 rows)
```

Let's create our first two gold tables, top_selling_items and sales_performance_24h

```
-- Top selling items by total revenue
CREATE TABLE gold.top_selling_items AS
SELECT
    item_id,
    item_name,
    item_category,
    SUM(total_price) AS total_revenue
FROM
    silver.purchases_enriched
GROUP BY
    item_id, item_name, item_category
ORDER BY
    total_revenue DESC
LIMIT 10;

-- Item sales revenue in the past 24 hours
CREATE TABLE gold.sales_performance_24h AS
SELECT
    p.purchase_hour AS purchase_hour,
    SUM(p.total_price) AS total_revenue
FROM
    silver.purchases_enriched p
```

```
WHERE
    p.created_at >= CURRENT_TIMESTAMP - INTERVAL '24' HOUR
GROUP BY
    purchase_hour
ORDER BY
    purchase_hour ASC;
```

We used CREATE TABLE AS SELECT (CTAS) queries to create both tables from a single silver table, purchases_enriched, which is already denormalized and enriched with item information. Derived fields like total_price and purchase_hour speed up the queries.

Next, we create the gold table, pageviews_by_channel from another denormalized silver table. This table returns a list of traffic sources by total page views.

```
CREATE TABLE gold.pageviews_by_channel AS
SELECT
    channel,
    COUNT(*) AS total_pageviews
FROM
    silver.pageviews_by_items
GROUP BY
    channel
ORDER BY
    total_pageviews DESC;
```

The final table, top_converting_items joins two silver tables to compute the items with the highest conversion rates.

```
CREATE TABLE gold.top_converting_items AS
SELECT
    pvi.item_id,
    pvi.item_name,
```

```sql
    pvi.item_category,
    COUNT(DISTINCT pvi.user_id) AS unique_pageview_users,
    COUNT(DISTINCT pe.user_id) AS unique_purchase_users,
    COUNT(pe.id) AS total_purchases,
    COUNT(pvi.user_id) AS total_pageviews,
    CASE
        WHEN COUNT(pvi.user_id) = 0 THEN 0
        ELSE CAST(COUNT(pe.id) AS DOUBLE) / COUNT(pvi.user_id)
    END AS conversion_rate
FROM
    silver.pageviews_by_items pvi
LEFT JOIN
    silver.purchases_enriched pe
    ON pvi.item_id = pe.item_id
    AND pvi.user_id = pe.user_id
    AND date(pvi.received_at) = pe.purchase_date
GROUP BY
    pvi.item_id,
    pvi.item_name,
    pvi.item_category
ORDER BY
    conversion_rate DESC
LIMIT 10;
```

The above code demonstrates how we join two silver tables, pageviews_by_items and purchases_enriched, to produce an aggregated gold table that computes the items with the highest conversion rates. This conversion rate represents the ratio of purchases to page views, indicating which items are most effective at converting customer interest into actual sales. The query also calculates additional metrics such as unique users viewing vs. purchasing, providing a comprehensive view of item performance.

CHAPTER 4   DATA VISUALIZATION WITH APACHE SUPERSET

Let's verify the gold tables created so far:

```
-- Let's check our newly created gold tables
trino:gold> show tables;
        Table
-----------------------
 pageviews_by_channel
 sales_performance_24h
 top_converting_items
 top_selling_items
(4 rows)
```

Now that we have successfully created and verified our gold tables, we're ready to move on to the next step: building dashboards in Superset that will visualize these key performance indicators.

Creating a BI Dashboard with Apache Superset

Now that we have created our gold tables with Trino, we're ready to move to the next phase of our data analytics journey – visualizing these metrics in an intuitive and interactive dashboard. While gold tables provide the structured, business-ready data we need, transforming these numbers into visual insights will help stakeholders quickly understand business performance and make data-driven decisions.

We'll use Apache Superset to create this dashboard. Superset is a modern, enterprise-ready business intelligence web application that allows users to create and share dashboards and explore data through intuitive visualizations. It has gained significant popularity in the data analytics space for several compelling reasons:

CHAPTER 4   DATA VISUALIZATION WITH APACHE SUPERSET

- Open-source and free: Unlike many commercial BI tools, Superset is completely open-source under the Apache license

- No-code visualization builder: Intuitive drag-and-drop interface makes it accessible to non-technical users

- Wide range of visualization types: Supports over 40 chart types, from basic bar charts to complex geospatial visualizations

- SQL Lab for direct querying: Allows data analysts to write custom SQL queries when needed

- Scalable architecture: Designed to handle large datasets and high concurrency

Superset connects directly to Trino using a database connection. This setup lets Superset use Trino's query engine to run SQL queries quickly across multiple data sources, including our Iceberg tables. The connection works through a SQLAlchemy driver – Superset sends SQL queries to Trino, Trino runs these queries on our gold tables, and then sends back results that Superset turns into visualizations.

For our OneShop analytics dashboard, this integration means we can create interactive visualizations directly from our gold tables without needing to move or duplicate data. Users can explore data through Superset's interface, with queries being processed by Trino in the background.

Before creating visualizations, let's create a database connection to Trino.

CHAPTER 4   DATA VISUALIZATION WITH APACHE SUPERSET

Access Superset by visiting http://localhost:8088/ in a browser. Log in using "admin" for both username and password. Click on **Settings ➤ Database Connections** on the top navigation to navigate to the database connections page. Then click on **+Database** to create a new connection.

You will notice that Trino appears in the supported databases list. That's because we've already installed the Trino SQLAlchemy driver into the Superset customer Docker image we launched initially. You can confirm it by checking ./superset/Dockerfile:

```
FROM apache/superset:latest

USER root

# Install Trino SQLAlchemy driver
RUN pip install 'trino[sqlalchemy]'

USER superset
```

Select "Trino" from the list and provide the following connection string for the SQLAlchemy driver. This will configure Superset to connect to Trino's Iceberg catalog.

```
trino://admin@trino:8080/iceberg
```

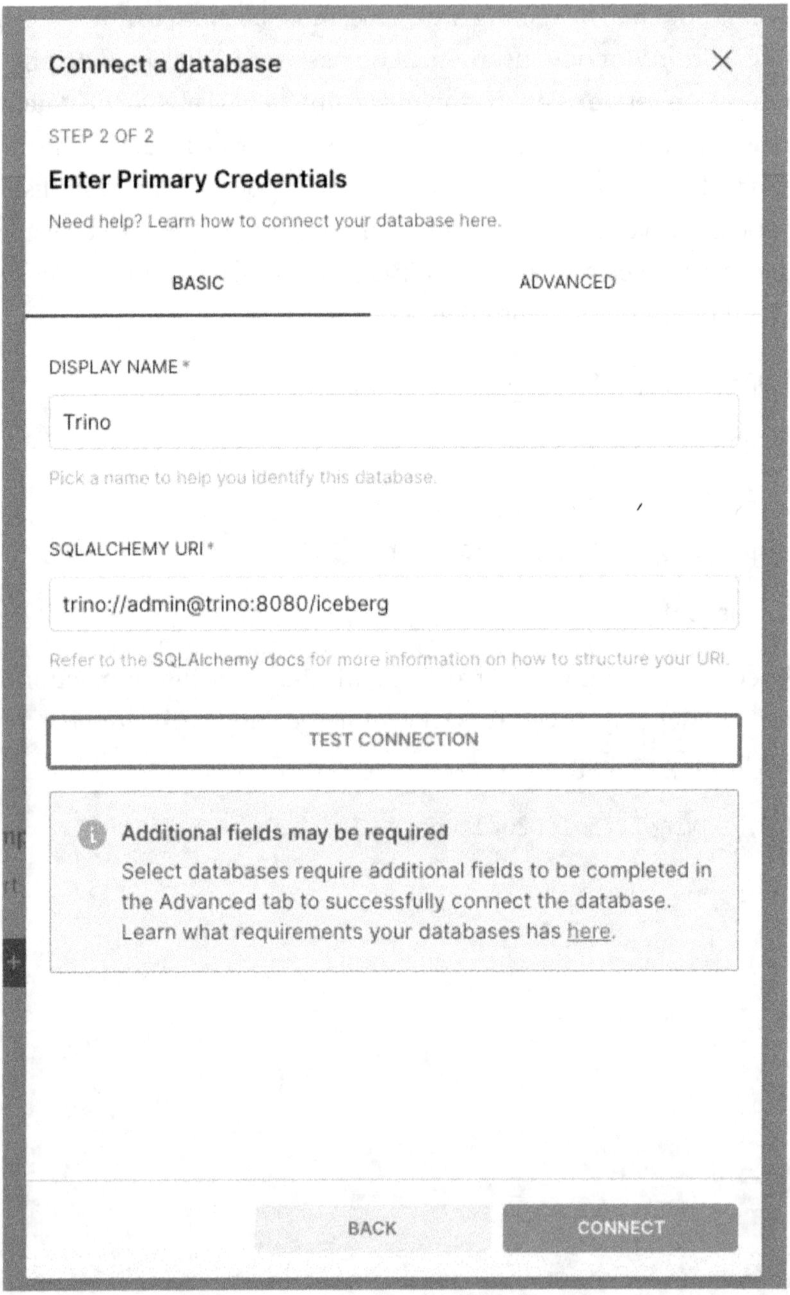

CHAPTER 4   DATA VISUALIZATION WITH APACHE SUPERSET

Figure 4-1. *Configure superset to connect to Trino's iceberg catalog*

CHAPTER 4   DATA VISUALIZATION WITH APACHE SUPERSET

Now, let's create our first visualization by creating a bar chart to visualize the past 24 hours' sales performance. This bar chart pulls data from the gold.sales_performance_24h Trino table.

Let's first explore the data using **SQL Lab**:

1. Navigate to the Superset interface and click on "SQL ➤ SQL Lab" in the top navigation bar

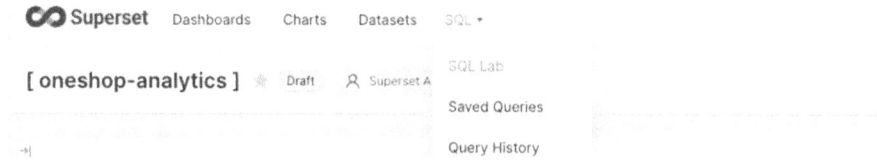

Figure 4-2. *SQL lab*

2. In the SQL editor, select **"Trino"** as the database from the dropdown menu

3. Enter the following query to examine the sales_performance_24h data:

```
SELECT * FROM gold.sales_performance_24h
ORDER BY purchase_hour ASC;
```

4. Click **"Run"** to execute the query and verify the data looks correct

5. Once satisfied with the query results, click **"Create Chart"** to visualize the data

6. Select **"Bar Chart"** as the visualization type

103

CHAPTER 4   DATA VISUALIZATION WITH APACHE SUPERSET

7. Configure the chart settings:

    - X-axis: purchase_hour

    - Y-axis: Select total_revenue field and then select SUM() from the Aggregate dropdown list.

    - Title: "Hourly Sales Performance"

8. Save the chart for now. We will add it later into a dashboard.

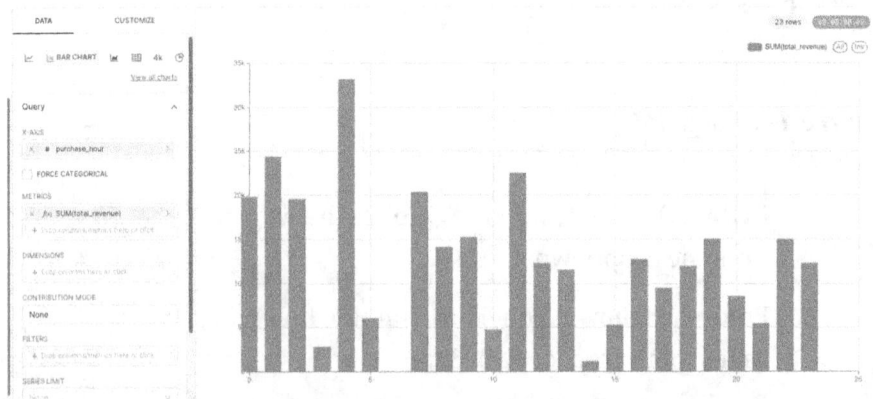

Figure 4-3. Hourly sales performance

Next, let's create a pie chart to visualize the breakdown of page views per traffic source.

1. Open a new tab in the SQL Lab

2. Select "**Trino**" as the database from the dropdown menu

3. Enter the following query to examine the gold.pageviews_by_channel data:

    SELECT * FROM gold.pageviews_by_channel;

CHAPTER 4   DATA VISUALIZATION WITH APACHE SUPERSET

4. Once satisfied with the query results, click "**Create Chart**" to visualize the data

5. Select "**Pie Chart**" as the visualization type

6. Configure the chart settings:
   - Dimension: channel
   - Metric: SUM(total_pageviews)
   - Title: "Page Views by Traffic Source"

7. Save the chart for now. We will add it later into a dashboard.

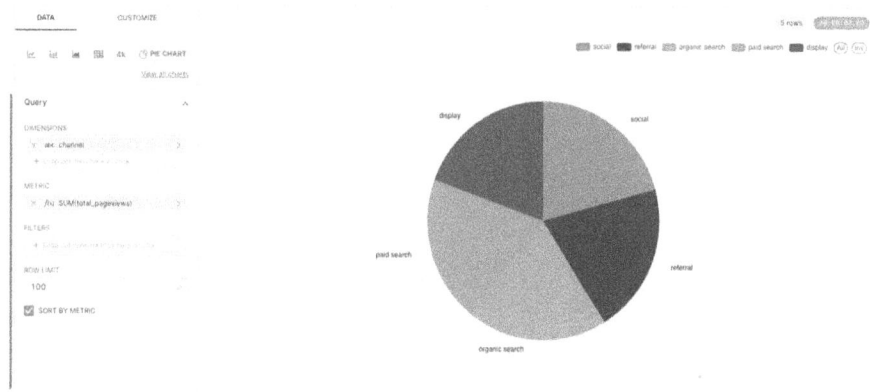

Figure 4-4. *Hourly sales performance pie-chart*

Once you have saved individual charts, Superset creates virtual datasets that can be reused across different visualizations. Let's create a dashboard to combine our charts:

1. Navigate to **Dashboards** in the top navigation bar

2. Click the + **Dashboard** button in the upper right corner

105

CHAPTER 4   DATA VISUALIZATION WITH APACHE SUPERSET

3. Enter a name for your dashboard (e.g., "*OneShop Analytics Dashboard*")

4. Click **Save**

5. In your new dashboard, click the **Edit Dashboard** button in the top right

6. Click on the + **Charts** tab in the left sidebar to see your saved charts

7. Drag and drop your "Hourly Sales Performance" bar chart onto the dashboard canvas

8. Similarly, drag and drop your "Page Views by Traffic Source" pie chart onto the dashboard

9. Arrange the charts by dragging and resizing them as needed

10. Click **Save** in the top right to save your dashboard layout

Your dashboard is now ready to view!

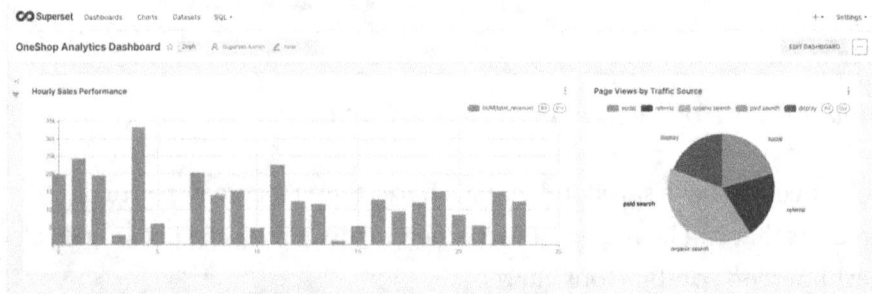

Figure 4-5. *Composite dashboard*

106

CHAPTER 4   DATA VISUALIZATION WITH APACHE SUPERSET

You can further enhance it by

- Adding filters to make the dashboard interactive
- Setting auto-refresh intervals to keep data current
- Adjusting chart properties for better visualization
- Adding text boxes to provide context or explanations

Note that we've only demonstrated how to create dashboards for two of the metrics we initially discussed. You can follow the same process to include the other KPIs we created in our gold tables, such as top selling items and top converting items. The final dashboard with all metrics would look like this.

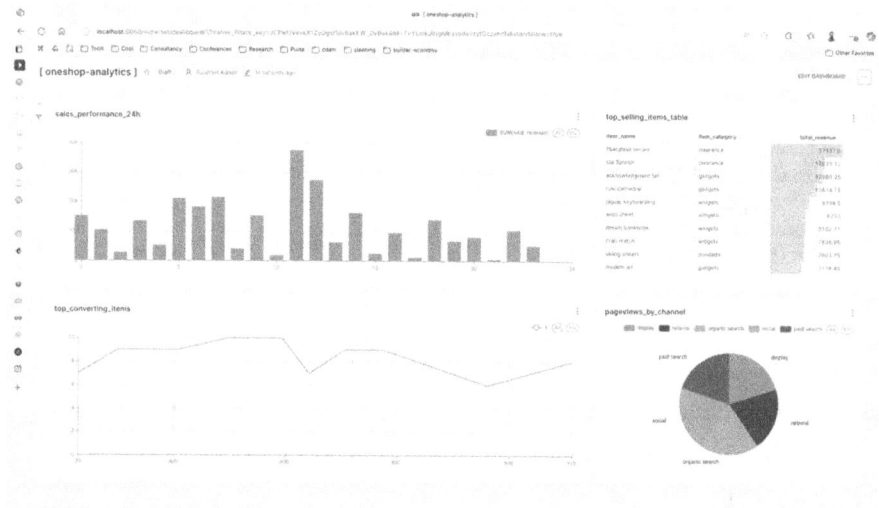

Figure 4-6. Enhanced dashboard

CHAPTER 4   DATA VISUALIZATION WITH APACHE SUPERSET

Summary

In this chapter, we explored building analytical tables in the gold layer and creating business intelligence dashboards with Trino and Apache Superset. Key points include:

- Using Trino to create gold layer tables that aggregate data for specific business metrics
- Creating five key business KPIs through SQL queries: sales performance, pageviews by channel, top selling items, and conversion rates
- Setting up Apache Superset to connect directly to Trino for visualization
- Building interactive dashboards that transform numerical data into visual insights
- Configuring different chart types (bar charts, pie charts) to represent various metrics
- Combining individual visualizations into a comprehensive business dashboard

This dashboard implementation completes the data pipeline work we started in **Chapter 3**, from raw data ingestion to deriving actionable business insights – enabling data-driven decision making for the OneShop team.

However, we performed several operations manually – including lakehouse hydration, Iceberg table creation, and data transformation. In a production environment, these processes need to be automated. In the next chapter, we will automate all pipeline operations with Apache Airflow.

CHAPTER 5

ETL Orchestration with Apache Airflow

The OneShop data engineering team has been on quite a journey. After successfully building their data lakehouse infrastructure in previous chapters, they now face a new challenge: automation. The team is tired of manually running scripts and processes every time they need to update their analytics. As Maya, the lead data engineer, puts it: "We're spending too much time babysitting our data pipelines when we could be building new features."

The immediate problem they're tackling involves the marketing team's customer segmentation needs. Currently, whenever the marketing team plans an outreach campaign, they request updated customer segments from the data team. This involves a data engineer manually running SQL queries to compute engagement metrics, exporting the results to CSV files, storing them in the team's shared MinIO bucket, and finally notifying the marketing team that the data is ready. This ad-hoc process is not only time-consuming but also prone to human error.

"What if we could set this up to run automatically every day?" suggests Alex, one of the data scientists. "The marketing team would always have fresh segments available, and we wouldn't have to drop everything to run these queries whenever they ask."

CHAPTER 5   ETL ORCHESTRATION WITH APACHE AIRFLOW

The team decides to implement a workflow orchestration solution using Apache Airflow to automate this entire process. Their goal is to create a pipeline that will

- Compute customer segments based on user engagement metrics stored in their lakehouse
- Export these segments to CSV files
- Store the files in a MinIO bucket accessible to the marketing team
- Send automatic email notifications to alert the marketing team when new segments are available

In this chapter, we'll follow the OneShop team as they implement this solution. We'll explore how to set up and configure Apache Airflow, create a DAG (Directed Acyclic Graph) to orchestrate the workflow, connect Airflow securely to Trino for executing SQL queries against their lakehouse, and implement the necessary tasks for file export and notifications. By the end of this chapter, you'll understand how to use Airflow to transform manual, ad-hoc data processes into automated, reliable pipelines that can run without human intervention.

Before diving into the technicalities of setting up our automated data pipeline, let's take a moment to understand the powerful tool at the center of this chapter: Apache Airflow.

Understanding Apache Airflow

Apache Airflow is an open-source platform designed to programmatically author, schedule, and monitor workflows. Created by Airbnb in 2014 to manage their increasingly complex workflows, it was later donated to the Apache Software Foundation and became a top-level Apache project in 2019. Today, it's one of the most popular workflow orchestration tools in the data engineering ecosystem.

At its core, Airflow allows data engineers to define workflows as code, track their execution, and handle dependencies between tasks. This "workflows as code" approach means that data pipelines can be version-controlled, tested, and maintained using software development best practices.

Airflow's Architecture and Components

Airflow's architecture consists of several key components working together:

- **Scheduler:** The heart of Airflow, responsible for triggering scheduled workflows and submitting tasks to the executor when their dependencies are complete.

- **Executor:** Determines how tasks are executed. Options include LocalExecutor (runs tasks on the same machine), CeleryExecutor (distributes tasks across worker nodes), and KubernetesExecutor (runs tasks as Kubernetes pods).

- **Webserver:** Provides a user-friendly UI for monitoring, triggering, and debugging workflows. It visualizes the DAGs, displays task status, and provides access to logs.

- **Workers:** Execute the tasks assigned by the executor. In distributed setups with CeleryExecutor, workers run on separate machines to process tasks in parallel.

- **Metadata Database:** Stores information about the state of tasks and DAGs. Typically uses PostgreSQL, MySQL, or SQLite to maintain execution history and status.

CHAPTER 5   ETL ORCHESTRATION WITH APACHE AIRFLOW

DAGs and Airflow Concepts

In Airflow, workflows are represented as Directed Acyclic Graphs (DAGs), which are collections of tasks with directional dependencies. DAGs define the order of execution and relationships between tasks, ensuring they run in the correct sequence.

Key concepts in Airflow DAGs include:

- **Tasks:** The individual units of work within a DAG. Each task is an instance of an operator and represents a single atomic job.

- **Operators: operators:** Templates for tasks that define what work gets done. Common examples include: PythonOperator (executes Python functions), BashOperator (runs Bash commands or scripts), EmailOperator (sends emails), and SQLOperator variants (execute SQL queries against various databases).

- **Task dependencies:** Define the execution order using the >> and << operators or the set_upstream() and set_downstream() methods.

- **Sensors:** Special operators that wait for certain conditions to be true, such as file existence or table updates.

- **XComs:** Allow tasks to exchange small amounts of data with each other.

For our lakehouse project, we'll be creating a DAG that orchestrates the processing of our user engagement data. The DAG will use the TrinoOperator, which allows Airflow to connect to and execute queries against our Trino server. This operator is installed during the Airflow image build process, ensuring it's available when our DAG runs.

CHAPTER 5 ETL ORCHESTRATION WITH APACHE AIRFLOW

The TrinoOperator is not part of Airflow's core operators but is available as a provider package. Our Dockerfile for Airflow includes the installation of the apache-airflow-providers-trino package, which adds the ability to connect to Trino servers and execute SQL queries.

```
FROM apache/airflow:2.8.4-python3.11

USER airflow
RUN pip install --no-cache-dir \
    apache-airflow-providers-trino \
    --constraint "<https://raw.githubusercontent.com/apache/
    airflow/constraints-2.8.4/constraints-3.11.txt>"
```

Starting Up the Lakehouse Components

Just as in the previous chapter, we need to ensure our lakehouse infrastructure is up and running. This will provide us with the silver tables necessary for computing customer segments later.

Navigate to the root level of the `chapter-05` directory and start all containers by running:

```
docker-compose up -d --build
```

This command builds the `loadgen` container and launches the entire Docker stack that we explored in **Chapter 3**. The stack includes preloaded Postgres and MinIO source systems, along with Trino and Mailhog containers. Mailhog is a fake SMTP server that we'll use later in this chapter to send email reminders.

Once started, run this script to start the lakehouse preparation process:

```
./lakehouse-preparer.sh
```

113

CHAPTER 5   ETL ORCHESTRATION WITH APACHE AIRFLOW

Once complete, our lakehouse will be fully populated with data flowing from the bronze through to the gold layer, ready for us to build customer segmentation tables in the gold layer.

Setting Up Apache Airflow Components

With the lakehouse infrastructure now operational, our next step is to set up Airflow.

We've separated the lakehouse and Airflow Docker components into two files for better organization and management. You can find the Airflow Docker Compose configuration in `airflow.yaml`.

```
x-airflow-common:
  &airflow-common
  build:
    context: ./airflow
  environment:
    AIRFLOW__CORE__EXECUTOR: LocalExecutor
    AIRFLOW__CORE__SQL_ALCHEMY_CONN: postgresql+psycopg2://airflow:airflow@airflow-db/airflow
    AIRFLOW__WEBSERVER__SECRET_KEY: "7KdlOpEf4Rvu14oVxRAZf3FVu1tE5bVpsbpYv7dGspM"
    AIRFLOW__WEBSERVER__EXPOSE_CONFIG: 'True'
    AIRFLOW__SMTP__SMTP_HOST: mailhog
    AIRFLOW__SMTP__SMTP_PORT: 1025
    AIRFLOW__SMTP__SMTP_MAIL_FROM: airflow@example.com
    AIRFLOW__SMTP__SMTP_STARTTLS: 'False'
    AIRFLOW__SMTP__SMTP_SSL: 'False'
    AIRFLOW__SMTP__SMTP_USER: ''
    AIRFLOW__SMTP__SMTP_PASSWORD: ''
```

```yaml
    volumes:
      - ./airflow/dags:/opt/airflow/dags
    networks:
        iceberg_net:
    depends_on:
      - airflow-db
services:
  airflow-db:
    container_name: airflow-db
    image: postgres:15
    environment:
      POSTGRES_USER: airflow
      POSTGRES_PASSWORD: airflow
      POSTGRES_DB: airflow
    volumes:
      - ./airflow/db:/var/lib/postgresql/data
    networks:
        iceberg_net:

  airflow-webserver:
    container_name: airflow-webserver
    <<: *airflow-common
    ports:
      - "8080:8080"
    command: webserver

  airflow-scheduler:
    container_name: airflow-scheduler
    <<: *airflow-common
    command: scheduler
```

```
  airflow-init:
    container_name: airflow-init
    <<: *airflow-common
    command: >
      bash -c "
        airflow db init &&
        airflow users create --username admin --password
        admin --firstname Admin --lastname User --role
        Admin --email admin@example.com
      "
networks:
  iceberg_net:
```

This Docker Compose configuration first defines shared settings for Airflow services using YAML anchors, including database connections, environment variables, and volume mappings.

- `airflow-db`: Runs PostgreSQL 15 as the metadata store for Airflow with persistent storage.

- `airflow-webserver`: Exposes the Airflow UI on port 8080 for monitoring and managing workflows.

- `airflow-scheduler`: Handles workflow scheduling and task distribution.

- `airflow-init`: One-time setup that initializes the database and creates an admin user.

Additionally, all Airflow services connect through the shared "iceberg_net" network, enabling communication between Airflow and other lakehouse components. This connectivity is essential for interactions between Airflow DAGs and Trino running inside the lakehouse, which we will discuss later.

CHAPTER 5   ETL ORCHESTRATION WITH APACHE AIRFLOW

The configuration uses LocalExecutor for Ariflow task execution and includes email notification settings through MailHog for development purposes. DAGs are mounted from the host machine for easy development. While we're using LocalExecutor here for simplicity, for production setups we recommend using CeleryExecutor or KubernetesExecutor which provide better scalability and distributed task execution capabilities.

Let's start our Airflow deployment by building the custom Airflow image and initializing the database.

First, run this command from the root level of chapter-05 to build the Airflow image with the TrinoOperator installed:

```
docker compose -f airflow.yaml build
```

Then initialize the Airflow database.

```
docker compose -f airflow.yaml up airflow-init
```

After running the above one-off command, several important steps occur automatically:

- **Database schema creation:** Airflow creates all necessary tables and relationships in the Postgres database (airflow-db). Other Airflow components will connect to this database via the SqlAlchemy connection URL specified in AIRFLOW__CORE__SQL_ALCHEMY_CONN parameter.

- **Admin user creation:** A default admin user (username: admin, password: admin) is created as specified in the initialization script.

- **Default connections setup:** Airflow creates basic connections required for its operation.

- **Configuration validation:** The system checks that your environment variables and settings are valid.

CHAPTER 5   ETL ORCHESTRATION WITH APACHE AIRFLOW

Once this initialization process completes successfully, you can proceed to start the main Airflow services with:

```
docker compose -f airflow.yaml up -d
```

This will start the Airflow webserver and scheduler in detached mode. After a few moments, you'll be able to access the Airflow UI by navigating to http://localhost:8080 in your web browser and logging in with the admin credentials created during initialization.

Once logged in, you will see a DAG file (user_engagement_segments_dag) has already been deployed under the DAGs page. Airflow automatically scans the mounted /opt/airflow/dags directory (mapped to your local ./airflow/dags folder) and registers any valid Python files containing DAG definitions.

Figure 5-1. Airflow automatically scans the mounted dags directory

We will break down the DAG's contents later.

Configuring TLS and HTTPS for Trino

Trino runs with no security by default. This allows you to connect to the server using URLs that specify the HTTP protocol when using the Trino CLI, the Web UI, or other clients. However, before we trigger the DAG, Trino must be configured to accept secure connections over HTTPS. This is necessary because the TrinoOperator in the DAG always connects to Trino over HTTPS.

CHAPTER 5 ETL ORCHESTRATION WITH APACHE AIRFLOW

To simplify the process, we've already completed the required configurations for you. However, we'll walk you through the relevant configuration files and settings to give you a better understanding.

For this project, we mounted an entire /etc directory structure into Trino container, which configures the required authentication settings. You will see the following folder structure inside the local ./trino/etc folder, which is mapped to /etc/trino inside the Trino container:

```
./trino/etc
├── catalog
│   └── iceberg.properties
├── config.properties
├── jvm.config
├── keystore.jks
├── node.properties
├── password-authenticator.properties
└── password.db
```

The most important files for TLS configuration are:

- config.properties: Contains the main Trino server configuration, including TLS settings.

- keystore.jks: The Java keystore containing the server's private key and certificate.

- password-authenticator.properties: Configures Trino's password-based authentication system.

- password.db: Contains the encrypted password credentials for users who can connect to Trino.

CHAPTER 5   ETL ORCHESTRATION WITH APACHE AIRFLOW

Let's look at the relevant sections in config.properties that enable TLS:

```
# Enable HTTPS connections on port 8443
http-server.https.enabled=true
http-server.https.port=8443

# Configures PASSWORD based authentication
http-server.authentication.type=PASSWORD

# Keystore configurations for the self-signed certificate
http-server.https.keystore.path=/etc/trino/keystore.jks
http-server.https.keystore.key=trino123

# Enable HTTP during development
http-server.authentication.allow-insecure-over-http=true

discovery.uri=https://localhost:8443
internal-communication.https.required=true
internal-communication.shared-secret=<A long string goes here>
```

The configuration line `http-server.authentication.type=PASSWORD` is setting up password-based authentication for Trino. This is Trino's implementation of HTTP basic authentication, where users must provide valid credentials to access the server.

The `password-authenticator.properties` file shown below specifies where Trino should find the password file, which is set to the `password.db` file.

```
password-authenticator.name=file
file.password-file=/etc/trino/password.db
```

The `password.db` file contains a list of usernames and passwords, one per line, separated by a colon. Passwords are securely hashed using bcrypt or PBKDF2 algorithms.

CHAPTER 5 ETL ORCHESTRATION WITH APACHE AIRFLOW

Running `cat password.db` reveals the username, `test`, and the encrypted value of password (`password` in this case).

`test:$2y$10$8e45bgBSFcw/beIYgTwuD.KetJNw7nKnX78ne9acBWi/GGmT2.hLG`

We will use these credentials when connecting to Trino in the Airflow DAG.

You can easily create a password file yourself by using the `htpasswd` command:

`htpasswd -B -C 10 password.db <your username>`

The setting `http-server.authentication.allow-insecure-over-http=true`, which (for development purposes) allows using this authentication over regular HTTP connections. In a production environment, you would typically set `allow-insecure-over-http` to false to ensure credentials are only transmitted over secure HTTPS connections.

Use the specified `keystore.jks` file with the password "trino123" for storing the server's TLS certificate and private key. This Java KeyStore (JKS) file provides the cryptographic materials needed to establish secure HTTPS connections and ensure encrypted communication between clients and the Trino server.

You can use a command similar to this to generate this file:

```
keytool -genkeypair \\
  -alias trino \\
  -keyalg RSA \\
  -keystore keystore.jks \\
  -storepass trino123 \\
  -keypass trino123 \\
  -dname "CN=localhost, OU=Test, O=Test, L=City, S=State, C=US"
```

The shared secret in internal-communication settings is essential for Trino cluster security, providing authentication between nodes, preventing unauthorized access, encrypting sensitive data exchange, and establishing trust across distributed components.

The shared secret should be a strong, random string and kept confidential, as anyone with access to this value could potentially authenticate as a legitimate Trino node. You can generate a large random string as shared secret by running:

```
openssl rand 512 | base64
```

The configuration also includes two other important security settings:

1. `discovery.uri=https://localhost:8443` ensures that the discovery service (which helps nodes find each other) operates over HTTPS instead of HTTP
2. `internal-communication.https.required=true` enforces that all inter-node communication must use HTTPS

Even though we're running a single-node setup in this example, configuring these security features is considered a best practice and would be essential if you later expand to a multi-node cluster.

To ensure Airflow can connect to our secure Trino instance, we need to configure the Trino connection in the Airflow UI. Follow these steps to set up the connection properly:

Configuring Trino Connection in Airflow

1. Open your web browser and navigate to the Airflow web interface at http://localhost:8080
2. Log in using the credentials we created during initialization (username: `admin`, password: `admin`)

CHAPTER 5   ETL ORCHESTRATION WITH APACHE AIRFLOW

3. In the top navigation menu, click on "Admin" and select "Connections" from the dropdown menu

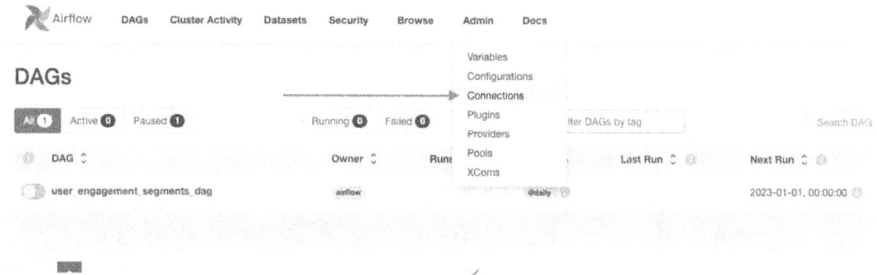

Figure 5-2. *Configuring trino connection in airflow*

4. Click the "+" button to add a new connection or find and edit the existing "trino_default" connection if it already exists 5. Configure the connection with the following parameters:

Connection Id:	trino_default
Connection Type:	Trino
Host:	trino
Port:	8443
Login:	test
Password:	password
Schema:	iceberg
Extra:	{"verify": false, "protocol": "https"}

123

The "Extra" field contains a JSON object with two important parameters:

- **verify:** Set to false to disable SSL certificate verification. This is useful in development environments with because we are using a self-signed certificate, but in production, you should set this to true and provide proper certificates.

- **protocol:** Set to "https" to ensure the connection uses HTTPS as required by our Trino configuration.

5. Click the "Save" button to store the connection settings

Once saved, Airflow will use these credentials and settings when executing the TrinoOperator tasks in our DAG. The connection uses the same username and password we configured in the Trino password.db file, and correctly specifies the HTTPS protocol required by our secure Trino setup.

Breaking Down the Airflow DAG File

Now that we have configured our Trino environment with the necessary security settings, let's turn our attention to the orchestration component of our data pipeline. We'll examine the Airflow DAG file that coordinates various operations within our lakehouse architecture.

In this section, we'll systematically break down each component of our Airflow DAG to understand how it orchestrates data processing workflows. We'll explore how the DAG connects to our secure Trino instance, executes SQL queries against our Iceberg tables, and manages the end-to-end data transformation process.

CHAPTER 5   ETL ORCHESTRATION WITH APACHE AIRFLOW

The complete DAG file is available in ./airflow/dags/user_ engagement_segments_dag.py. Since this file is quite long, we'll focus only on the most important sections of the code.

The following code defines the DAG by giving it an ID, setting its start date, and establishing a daily schedule interval—meaning this workflow will execute once every day.

```
with DAG(
    dag_id="user_engagement_segments_dag",
    start_date=datetime(2023,1,1),
    schedule_interval='@daily'
) as dag:
```

Connecting the DAG to Trino with Trino Operator

The DAG's first task uses the TrinoOperator to establish a secure connection to the Trino server and execute a SQL query that computes customer segments based on user engagement metrics. This is the core analytical component of our DAG.

```
segment_users = TrinoOperator(
        task_id='segment_users',
        sql='sql/trino.sql',
        trino_conn_id='trino_default'
)
```

The TrinoOperator connects to the Trino container in our lakehouse infrastructure using the secure trino_default connection ID we configured earlier.

CHAPTER 5 ETL ORCHESTRATION WITH APACHE AIRFLOW

After establishing the connection, this task runs the SQL code below, which is loaded from the ./airflow/dags/sql/trino.sql file:

```sql
CREATE OR REPLACE TABLE iceberg.gold.user_engagement_segments
AS
SELECT
    u.id AS user_id,
    u.email,
    u.full_name,
    COUNT(p.page) AS total_pageviews,
    COUNT(DISTINCT DATE(p.received_at)) AS active_days,
    MAX(DATE(p.received_at)) AS last_active_date,
    DATE_DIFF('day', MAX(DATE(p.received_at)), CURRENT_DATE) AS
    days_since_last_active,
    CASE
        WHEN COUNT(p.page) >= 50 AND DATE_DIFF('day',
        MAX(DATE(p.received_at)), CURRENT_DATE) <= 3 THEN
        'high_engagement'
        WHEN COUNT(p.page) BETWEEN 10 AND 49 AND DATE_DIFF
        ('day', MAX(DATE(p.received_at)), CURRENT_DATE) <= 7
        THEN 'medium_engagement'
        ELSE 'low_engagement'
    END AS engagement_segment
FROM iceberg.silver.users u
LEFT JOIN iceberg.silver.pageviews_by_items p ON u.id =
p.user_id
WHERE u.valid_email = TRUE
GROUP BY u.id, u.email, u.full_name
```

CHAPTER 5   ETL ORCHESTRATION WITH APACHE AIRFLOW

In summary, this SQL query creates or replaces a table called user_engagement_segments in the Iceberg gold layer to store segmented customers. First, the user engagement is analyzed by joining two silver tables: users and pageviews_by_items ,calculating metrics such as total page views, active days, and days since last activity.

Based on these metrics, the query segments users into three engagement categories:

- **High engagement**: Users with 50+ page views who were active within the last 3 days
- **Medium engagement**: Users with 10-49 page views who were active within the last week
- **Low engagement**: All other users

Finally, the query filters for users with valid email addresses and groups results by user identifiers.

Using the "CREATE OR REPLACE TABLE" approach in a daily scheduled job has several practical implications. This approach completely rebuilds the table each time the DAG runs, which means all historical engagement data is recalculated with each execution.

For smaller datasets or early-stage implementations, this approach is often practical despite the inefficiency. However, as data volumes grow, you might consider alternative approaches like incremental processing (only processing new or changed records since the last run), partitioning (organizing the table by date partitions and only replacing recent partitions), or merge operations (using "MERGE INTO" statements to update existing records and insert new ones).

Given that this query categorizes user engagement based on recent activity, the full rebuild approach ensures that engagement segments remain current and accurate, which may justify the resource cost of recreating the table daily.

CHAPTER 5   ETL ORCHESTRATION WITH APACHE AIRFLOW

Exporting Customer Segments into MinIO as CSV

Next two tasks use the Python operator to run two Python functions defined in the DAG file.

```
export_csv = PythonOperator(
    task_id='export_to_csv',
    python_callable=export_segmented_users_to_csv,
    provide_context=True,
)

upload_to_minio = PythonOperator(
    task_id='upload_to_minio',
    python_callable=upload_csv_to_minio,
    provide_context=True,
)
```

The first task, export_csv, invokes the export_segmented_users_to_csv() function to query the gold.user_engagement_segments table and save the results as a CSV file. The second task, upload_to_minio, calls the upload_csv_to_minio() function to store this CSV file in a MinIO bucket. Each file is named with the execution date appended. We've omitted the implementation details of these functions for brevity.

Notifying the Success of the Operation

After computing the customer segments and exporting them to MinIO, the final task sends an email notification to the marketing team confirming the successful operation.

```
notify_success = EmailOperator(
    task_id='notify_success',
    to='marketing-team@example.com',
    subject='[Airflow] User Engagement Segments Exported',
```

```
    html_content="""
        <p>Hello Team,</p>
        <p>The user engagement segments have been refreshed
        and exported to a CSV file.</p>
        <p>File location: <code>/tmp/segmented_users_{{
        ds }}.csv</code></p>
        <p>- Airflow Bot</p>
    """
)
```

This notification enables the marketing team to import the CSV file into their CRM or marketing automation tool for planning outreach campaigns.

For this email functionality, we implement the built-in EmailOperator with MailHog serving as our development email server. MailHog is a valuable tool for development and testing environments as it provides a simple way to capture and inspect outgoing emails without actually sending them to real recipients. This is particularly useful when testing notification workflows like the one in our DAG.

This Email operator utilizes the environment variables that we configured in the `airflow.yaml` file and applied to all Airflow containers.

```
AIRFLOW__SMTP__SMTP_HOST: mailhog
    AIRFLOW__SMTP__SMTP_PORT: 1025
    AIRFLOW__SMTP__SMTP_MAIL_FROM: airflow@example.com
    AIRFLOW__SMTP__SMTP_STARTTLS: 'False'
    AIRFLOW__SMTP__SMTP_SSL: 'False'
    AIRFLOW__SMTP__SMTP_USER: ''
    AIRFLOW__SMTP__SMTP_PASSWORD: ''
```

While MailHog works perfectly for our development setup, in a production environment you would need to configure Airflow to use a real SMTP server.

CHAPTER 5   ETL ORCHESTRATION WITH APACHE AIRFLOW

Finally, all tasks are sequenced together using Airflow's task dependency system, creating a linear execution flow:

```
segment_users >> export_csv >> upload_to_minio >> notify_success
```

This dependency chain ensures that tasks execute in the proper order: first, the TrinoOperator segments the users, then the data is exported to a CSV file, followed by uploading this file to MinIO, and finally sending a notification email. These dependencies are critical because each task relies on the successful completion of its predecessor. When the DAG runs, the Airflow scheduler executes these tasks sequentially, monitoring each step and only proceeding to the next task when the previous one completes successfully. If any task fails, the execution stops at that point, preventing downstream tasks from running with invalid or missing data, thus maintaining the integrity of our data pipeline.

Triggering the DAG

Now that our DAG file is ready, let's see how to trigger it manually in the Airflow UI and verify that all components of our data pipeline are working as expected.

While our DAG is scheduled to run automatically on a daily basis in production, during development and testing phases, we can trigger it manually to verify its functionality:

1. Navigate to the Airflow UI in your browser at http://localhost:8080

2. Log in with the default credentials (username: airflow, password: airflow)

3. From the DAGs list, locate the *"user_engagement_segments_dag"*

CHAPTER 5   ETL ORCHESTRATION WITH APACHE AIRFLOW

4. Click on the play button (▶) icon on the right side of the DAG row, then select "Trigger DAG" from the dropdown menu

5. Optionally, you can specify a custom execution date or add configuration parameters in the dialog that appears

6. Click "Trigger" to start the DAG execution

The DAG will begin execution immediately, and you can monitor its progress in the Airflow UI. Each task will transition from "queued" to "running" to "success" (or "failed" if there are issues).

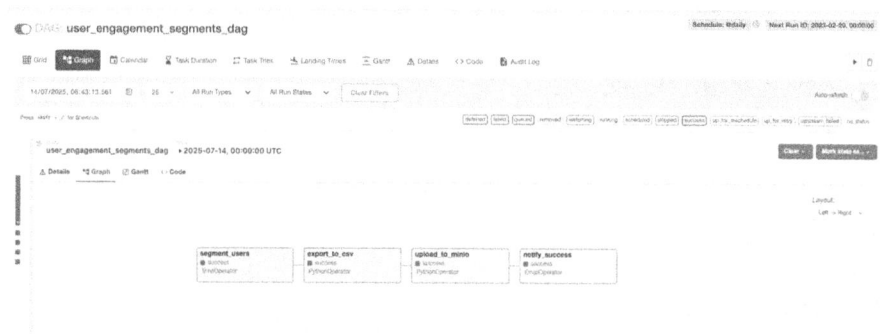

Figure 5-3. DAG execution

Remember that in production, this DAG will run automatically according to its schedule_interval configuration (@daily), so you won't need to trigger it manually. The scheduler component of Airflow will handle this based on the start_date and schedule_interval parameters defined in the DAG.

CHAPTER 5   ETL ORCHESTRATION WITH APACHE AIRFLOW

Verifying the Results of the DAG Execution

After triggering the DAG, we should verify that all components worked correctly by checking various parts of our lakehouse system:

First, verify the task run status in Airflow by navigating to the "Grid" view of the DAG to see a visual representation of the task execution. From there, click on each task to view its logs and verify successful execution. Make sure all tasks show a "success" status, which is typically indicated by a green color in the Airflow UI.

Next, we can verify whether the corresponding table as been created in Trino's gold layer.

```
docker compose exec trino trino

-- Switch to gold layer
use iceberg.gold;

-- Inspect the new user engagement segments table
SELECT
  engagement_segment,
  COUNT(*) as user_count
FROM iceberg.gold.user_engagement_segments
GROUP BY engagement_segment;
```

The above query should return counts of users in each engagement segment, confirming that the segmentation logic worked correctly.

You can also check the `customer-segments` MinIO bucket as a confirmation. Navigate to the MinIO web console at `http://localhost:9001`, log in with the configured credentials, and browse to the bucket where the CSV file was uploaded. Verify that a file named `segmented_users_YYYY-MM-DD.csv` exists (where YYYY-MM-DD corresponds to the execution date) and download the file to confirm it contains the expected user segment data.

CHAPTER 5   ETL ORCHESTRATION WITH APACHE AIRFLOW

Finally, check whether the email has been sent by:

- Access the MailHog web interface at http://localhost:8025.

- Verify that an email with the subject "*[Airflow] User Engagement Segments Exported*" was received

- Open the email to confirm it contains the expected content about the exported segments

Figure 5-4. Email verification

By verifying each of these components, you can ensure that your entire data pipeline is functioning correctly – from data processing in Trino to file export to MinIO, and finally to notification delivery.

If any issues are detected during verification, you can examine the logs in Airflow UI for the specific task that failed, make the necessary adjustments to your DAG or infrastructure configuration, and then trigger the DAG again to test your fixes.

Cleaning Up the Environment

After experimenting with the Airflow deployment, you can clean up your environment to free up system resources. The OneShop team has provided a cleanup script that will stop all containers and remove any resources created during this chapter's exercises.

CHAPTER 5 ETL ORCHESTRATION WITH APACHE AIRFLOW

To clean up your environment, run the following command from the project root directory:

```
./cleanup.sh
```

This script will:

- Stop all running Docker containers from the compose deployment
- Remove any associated volumes to free up disk space
- Remove any created networks

You can always restart the environment later by running the Docker Compose commands described earlier in this chapter if you want to continue experimenting with Airflow and the data pipeline.

Summary

In this chapter, we explored how the OneShop data engineering team tackled automation challenges by implementing Apache Airflow for their data workflows. We learned how to build a complete data pipeline using Airflow DAGs to automate customer segmentation, configure task dependencies to ensure proper execution order, use specialized operators like TrinoOperator and PythonOperator to process data, implement notification systems using EmailOperator, and test and verify pipeline execution through the Airflow UI.

With this foundation in place, the OneShop team can extend their automation capabilities in several ways. They can create additional DAGs for other recurring data tasks, such as data quality checks, model training, or dashboard refreshes. They can implement more complex workflow patterns using branching, dynamic task generation, or parallel execution. Adding error handling and retry mechanisms will make pipelines more robust, while setting up sensors can trigger workflows based on external events like file arrivals or API responses. Finally, integration with CI/CD systems will automate the deployment of new or updated DAGs.

The business impact of implementing Airflow has been significant for OneShop. The marketing team now receives daily updated customer segments without manual intervention from the data team. This automation has reduced the time-to-insight from days to hours, allowing marketing to launch more timely campaigns. It has freed up the data engineering team to focus on higher-value work instead of repetitive manual tasks. The solution has improved data quality and consistency by eliminating human error in the execution process, created a scalable foundation that can grow alongside OneShop's data needs, and enabled more data-driven decision-making across the organization by making reliable data available on a consistent schedule.

By the end of this implementation, Maya and her team transformed what was once a time-consuming manual process into an efficient, automated workflow – a crucial step in OneShop's journey toward becoming a truly data-driven organization.

PART II

Streaming Data and Real-Time Analytics

CHAPTER 6

Real-time Change Data Capture with Kafka and Debezium

Introduction

With the successful implementation of batch processing and data visualizations, OneShop's data team faced a new challenge: responding to customer events in real-time. Maya, the lead data engineer, noticed a significant issue with their inventory system. "Our customers are getting frustrated when they order products that show as available on our website, only to find out later that these items are actually out of stock," she explained during a team meeting.

The product team confirmed this problem: the current system only updates inventory data in batches, running every few hours. This delay meant customers sometimes ordered items that had already sold out, resulting in cancellations and disappointed shoppers.

"What if we could update our inventory information in real-time?" suggested Alex, one of the data engineers. "When an item sells out, we could immediately reflect that on the website."

CHAPTER 6   REAL-TIME CHANGE DATA CAPTURE WITH KAFKA AND DEBEZIUM

In this chapter, we'll build a robust streaming architecture that captures live data changes from source data systems. We'll start by implementing a real-time change data feed from the items table in Postgres, using Change Data Capture (CDC) with Debezium to stream inventory updates directly to Kafka topics.

We will then set up an OpenSearch index that acts as a real-time inventory dashboard. Through Kafka Connect's source and sink connectors, we'll create a seamless data pipeline between Postgres and OpenSearch. Along the way, we'll use Kafka Connect's Single Message Transforms (SMTs) to transform the captured change events into a format optimized for OpenSearch, ensuring efficient search capabilities while maintaining near real-time updates of inventory changes.

By the end of this hands-on chapter, you will learn practical skills about

- Building event streaming pipelines with Apache Kafka
- Implementing CDC patterns using Debezium
- Configuring and managing Kafka Connect, using SMTs
- Full-text search with OpenSearch

Before diving into our solution, let's examine the fundamentals of streaming data, Apache Kafka, and its ecosystem.

Streaming Data, Apache Kafka, and Kafka Connect

Streaming data represents a fundamental shift from traditional batch processing. In batch processing, data is collected over time and processed in large chunks at scheduled intervals. In contrast, streaming data is processed continuously as it arrives, enabling real-time analysis and immediate action. This approach is particularly valuable for use cases that require instant insights or responses, such as fraud detection, real-time analytics, or monitoring systems.

The benefits of stream processing include reduced latency, as data is processed immediately rather than waiting for the next batch window, and the ability to handle infinite streams of data. It also enables real-time decision making and more responsive business operations.

Apache Kafka, originally developed by LinkedIn and later open-sourced, has become the de facto standard for handling streaming data. At its core, Kafka is a distributed event streaming platform that can handle trillions of events per day. It's built around a few key concepts:

- **Topics:** Named channels that store streams of related events
- **Producers:** Applications that publish events to Kafka topics
- **Consumers:** Applications that subscribe to topics and process the events
- **Brokers:** Servers that form the Kafka cluster and store the events

Kafka's architecture ensures high throughput, fault tolerance, and horizontal scalability, making it ideal for building real-time data pipelines.

Speaking of data pipelines, Kafka Connect is a framework for building scalable, reliable streaming pipelines between Kafka and other systems. It provides a standardized way to integrate Kafka with external systems, whether as sources (bringing data into Kafka) or sinks (sending data out to other systems). One of its powerful features is Single Message Transforms (SMTs), which allow you to modify messages as they flow through the pipeline. Later in this chapter, we'll use SMTs to transform our CDC events into a format that's optimized for OpenSearch.

Change Data Capture and Debezium

CDC is a design pattern that identifies and captures changes made to data in a database, then ensures those changes are replicated to other systems in real-time. Instead of periodically scanning entire tables for changes, CDC tracks modifications at the source – including inserts, updates, and deletes – as they happen.

CDC works by monitoring the database's transaction logs or similar change tracking mechanisms. When a change occurs, CDC captures essential details, including the type of operation (insert, update, or delete), the modified data, timestamp, and relevant metadata about the change.

Debezium is an open-source distributed platform that turns existing databases into event streams, enabling applications to detect and respond to row-level changes in real-time. It's built on top of Apache Kafka and provides connectors for various databases, including PostgreSQL, MySQL, MongoDB, and others.

As a robust CDC solution, Debezium handles complex scenarios like schema changes, large transactions, and failure recovery. It's particularly valuable in microservices architectures where different services need consistent and up-to-date data.

For OneShop, Debezium serves as a vital tool that watches for any changes in its inventory database. When stock levels change in Postgres, Debezium instantly picks up these changes and sends them to Kafka. This means customers always see the correct stock numbers when shopping, which helps prevent issues like overselling products.

Before You Begin

You will find the code for this chapter located in the chapter-06 folder. If you haven't set everything up yet, refer to the **Prerequisites** section in **Chapter 1** for more information.

CHAPTER 6 REAL-TIME CHANGE DATA CAPTURE WITH KAFKA AND DEBEZIUM

Navigate to the project folder on a terminal:

cd <repository_root>/chapter-06

In this folder, you will find everything you need to deploy the inventory CDC pipeline we discussed in this chapter.

Breakdown of the Docker-Compose File

The chapter-06/docker-compose.yml file defines the services that make up our CDC pipeline.

```
services:
  zookeeper:
    container_name: zookeeper
    image: bitnami/zookeeper:3.8
    environment:
      ALLOW_ANONYMOUS_LOGIN: "yes"
    ports:
      - "2181:2181"

  kafka:
    container_name: kafka
    image: bitnami/kafka:3.6
    depends_on:
      - zookeeper
    ports:
      - "9092:9092"
    environment:
      KAFKA_CFG_ZOOKEEPER_CONNECT: zookeeper:2181
      KAFKA_CFG_LISTENERS: PLAINTEXT://:9092
      KAFKA_CFG_ADVERTISED_LISTENERS: PLAINTEXT://kafka:9092
      KAFKA_CFG_AUTO_CREATE_TOPICS_ENABLE: "true"
      ALLOW_PLAINTEXT_LISTENER: "yes"
```

CHAPTER 6   REAL-TIME CHANGE DATA CAPTURE WITH KAFKA AND DEBEZIUM

```
postgres:
  container_name: postgres
  hostname: postgres
  image: debezium/postgres:17
  ports:
    - "5432:5432"
  environment:
    POSTGRES_USER: postgresuser
    POSTGRES_PASSWORD: postgrespw
    POSTGRES_DB: oneshop
  volumes:
    -./postgres/postgres_bootstrap.sql:/docker-entrypoint-
    initdb.d/postgres_bootstrap.sql
  healthcheck:
    test: ["CMD-SHELL", "pg_isready -U postgresuser -d
    oneshop"]
    interval: 1s
    start_period: 60s

connect:
  container_name: connect
  build:
    context: ./kafka-connect
  depends_on:
    - kafka
    - postgres
  ports:
    - "8083:8083"
  environment:
    BOOTSTRAP_SERVERS: kafka:9092
    GROUP_ID: 1
    CONFIG_STORAGE_TOPIC: connect-configs
```

```yaml
      OFFSET_STORAGE_TOPIC: connect-offsets
      STATUS_STORAGE_TOPIC: connect-status
      KEY_CONVERTER: org.apache.kafka.connect.json.
      JsonConverter
      VALUE_CONVERTER: org.apache.kafka.connect.json.
      JsonConverter
      INTERNAL_KEY_CONVERTER: org.apache.kafka.connect.json.
      JsonConverter
      INTERNAL_VALUE_CONVERTER: org.apache.kafka.connect.json.
      JsonConverter
      CONNECT_REST_ADVERTISED_HOST_NAME: connect
      PLUGIN_PATH: /kafka/connect
opensearch:
  image: opensearchproject/opensearch:2.11.0
  container_name: opensearch
  environment:
    - cluster.name=opensearch-cluster
    - node.name=opensearch
    - discovery.type=single-node
    - bootstrap.memory_lock=true
    - "OPENSEARCH_JAVA_OPTS=-Xms512m -Xmx512m"
    - "DISABLE_INSTALL_DEMO_CONFIG=true"
    - "DISABLE_SECURITY_PLUGIN=true"
  ulimits:
    memlock:
      soft: -1
      hard: -1
    nofile:
      soft: 65536
      hard: 65536
```

```
    ports:
      - "9200:9200"
      - "9600:9600"
  items-loadgen:
    build: items-loadgen
    container_name: items-loadgen
    init: true
    depends_on:
      postgres: {condition: service_healthy}
  redpanda-console:
    image: docker.redpanda.com/redpandadata/console:latest
    container_name: redpanda-console
    ports:
      - "8080:8080"
    environment:
      KAFKA_BROKERS: kafka:9092
```

Here's a breakdown of each service:

- `postgres` – The source Postgres database containing the `items` table.

- `kafka` – A Kafka cluster with a single broker.

- `zookeeper` – Apache ZooKeeper, a distributed coordination service that maintains configuration data, naming, and synchronization across Kafka's distributed cluster nodes.

- `connect` – Kafka Connect with Debezium and OpenSearch connectors pre-installed.

- `opensearch` – A single-node OpenSearch cluster.

CHAPTER 6   REAL-TIME CHANGE DATA CAPTURE WITH KAFKA AND DEBEZIUM

- `redpanda-console` – Redpanda Console is a graphical user interface (GUI) for managing Kafka. We'll use this to browse Kafka topics later.
- `items-loadgen` – A Python script that populates the `items` table with 1,000 random items.

Running the Setup

From the root level of the `chapter-06` directory, start all containers by running:

```
docker compose up -d --build
```

This command will first build the `connect` and `items-loadgen` containers, then start all containers in the background. While this demo uses single containers for most services, a production environment would require a distributed deployment with multiple container instances. However, this setup provides a convenient local environment to learn the basics of a real-time CDC pipeline.

Source Database and Schema

We'll use the same `oneshop` schema from **Chapter 3**. For this chapter, we'll focus only on the `items` table since it tracks inventory levels for each product.

The initialization script `./chapter-06/postgres/postgres_bootstrap.sql` creates the `items` table with the following schema.

```
CREATE TABLE IF NOT EXISTS items
(
    id SERIAL PRIMARY KEY,
    name VARCHAR(100),
```

```
    category VARCHAR(100),
    price DECIMAL(7,2),
    inventory INT,
    inventory_updated_at TIMESTAMP DEFAULT CURRENT_TIMESTAMP,
    created_at TIMESTAMP DEFAULT CURRENT_TIMESTAMP,
    updated_at TIMESTAMP DEFAULT CURRENT_TIMESTAMP
);
```

Notice the `inventory` column. This is where the inventory level of an item is tracked.

Login to the Postgres container to check the table's contents:

```
docker compose exec postgres psql -U postgresuser -d oneshop
```

The `items` table is in the `public` schema of the `oneshop` database. It has been pre-populated with 1,000 items by a Python script in the `items-loadgen` container.

You can examine the first 10 records to get an idea.

```
oneshop=# SELECT * FROM public.items LIMIT 10;
```

Change Feed Creation

Our first task is to create a change data feed for the `items` table. We do this by creating a Debezium source connector configuration inside Kafka Connect.

For convenience, we're using a pre-built Kafka Connect image that already includes the Debezium connector plugins. This saves us from the manual process of downloading and installing these connectors. The Dockerfile for this custom image can be found in the `kafka-connect` directory.

CHAPTER 6 REAL-TIME CHANGE DATA CAPTURE WITH KAFKA AND DEBEZIUM

Exit the Postgres shell by typing exit. Then run the following command to create the source connector:

```
docker compose exec connect curl -i -X POST <http://localhost:
8083/connectors> \\
  -H "Content-Type: application/json" \\
  -d '{
    "name": "cdc-connector",
    "config": {
      "connector.class": "io.debezium.connector.postgresql.
      PostgresConnector",
      "database.hostname": "postgres",
      "database.port": "5432",
      "database.user": "postgresuser",
      "database.password": "postgrespw",
      "database.dbname": "oneshop",
      "database.server.name": "oneshop",
      "table.include.list": "public.items",
      "topic.prefix": "dbz",
      "plugin.name": "pgoutput",
      "slot.name": "cdc_slot",
      "publication.name": "cdc_pub",
      "decimal.handling.mode": "double",
      "key.converter": "org.apache.kafka.connect.json.
      JsonConverter",
      "value.converter": "org.apache.kafka.connect.json.
      JsonConverter"
    }
  }'
```

Let's break down the key components of this connector configuration:

- **connector.class:** Specifies the Debezium Postgres connector, which is responsible for capturing changes from Postgres databases

- *database. parameters**: Define the connection details for the Postgres database, including hostname, port, credentials, and database name. These parameters tell Debezium where to find the source database

- **table.include.list**: Specifies which tables to monitor for changes. In this case, we're only monitoring the "public.items" table

- ***.converter**: The *key.converter* and *value.converter* settings use JSON format for both the record key and value, making it easier to work with the data downstream

- **plugin.name**: Set to "*pgoutput*", which is Postgres's native logical replication output plugin. This plugin allows Debezium to capture changes using Postgres's native replication protocol, providing better performance and reliability compared to older methods.

- **decimal.handling.mode**: Set to "double" to handle decimal numbers appropriately when converting from Postgres's decimal type to Kafka's format. Otherwise, you'd see that they are formatted as random strings.

If the connector creation is successful, you should see a response with HTTP status code 201. If you receive a different status code, it's likely due to a malformed connector creation request.

```
HTTP/1.1 201 Created
Date: Wed, 04 Jun 2025 09:41:54 GMT
Location: <http://localhost:8083/connectors/cdc-connector>
Content-Type: application/json
Content-Length: 965
Server: Jetty(9.4.52.v20230823)

<JSON-formatted connector configuration goes here>
```

Once the source connector is deployed, it automatically creates the Kafka topic named `dbz.public.items`. The topic naming logic is straightforward: Debezium combines the topic prefix `dbz` (which we provided in the configuration) with the table name.

You can verify the topic exists in Kafka by running:

```
docker compose exec kafka kafka-topics.sh --bootstrap-server localhost:9092 --list
```

You should see the topic name in the output. If you don't see the topic, something has gone wrong. You can troubleshoot by checking the Kafka Connect container logs by running:

```
docker compose logs -f connect
```

When the Debezium connector first starts up, it performs what's called a "snapshot" of the source table. This initial snapshot captures the current state of the entire table and loads it into the Kafka topic. That's why you'll see all 1,000 records from our `items` table appear in the topic initially. After this snapshot is complete, Debezium switches to streaming mode where it only captures and publishes new changes as they occur.

CHAPTER 6   REAL-TIME CHANGE DATA CAPTURE WITH KAFKA AND DEBEZIUM

Let's run a Kafka Console Consumer to check the first five messages in the Kafka topic:

```
docker compose exec kafka kafka-console-consumer.sh \\
  --bootstrap-server localhost:9092 \\
  --topic dbz.public.items \\
  --max-messages 5 \\
  --from-beginning
```

This will result in a few JSON formatted messages like this:

```
{"id":1,"name":"July halibut","category":"doodads","price":289.96,"inventory":4914,"inventory_updated_at":1749029718949442,"created_at":1749029718949442,"updated_at":1749029718949442}
{"id":2,"name":"television wash","category":"gadgets","price":155.97,"inventory":4170,"inventory_updated_at":1749029718949442,"created_at":1749029718949442,"updated_at":1749029718949442}
{"id":3,"name":"signature discovery","category":"doodads","price":199.24,"inventory":2285,"inventory_updated_at":1749029718949442,"created_at":1749029718949442,"updated_at":1749029718949442}
{"id":4,"name":"receipt handicap","category":"clearance","price":473.72,"inventory":811,"inventory_updated_at":1749029718949442,"created_at":1749029718949442,"updated_at":1749029718949442}
{"id":5,"name":"wren passenger","category":"clearance","price":22.38,"inventory":1379,"inventory_updated_at":1749029718949442,"created_at":1749029718949442,"updated_at":1749029718949442}
```

Processed a total of five messages

CHAPTER 6   REAL-TIME CHANGE DATA CAPTURE WITH KAFKA AND DEBEZIUM

Each message in the topic represents a change event from the items table. The message value contains both the change data ('after' state) and metadata about the change event.

While the Kafka CLI tools are great for debugging in the terminal, a graphical interface can make it easier to browse and inspect messages. Redpanda Console provides a user-friendly web interface for exploring Kafka topics and messages.

To access the Redpanda Console, open your web browser and navigate to:

<http://localhost:8080>

In the Redpanda Console:

1. Click on "**Topics**" in the left sidebar
2. Find and click on the dbz.public.items topic
3. Click on the "**Messages**" tab to see the messages in the topic

The console provides a clean JSON viewer that makes it easy to inspect message keys, values, headers, and timestamps. You can also filter messages, search for specific content, and view message metadata – features that are more cumbersome to use with CLI tools.

Transforming the Change Event Structure

Let's look at the original message structure from Debezium:

```
{
  "schema": { ... },
  "payload": {
    "before": null,
    "after": {
```

```
    "id": 1,
    "name": "Gaming Mouse",
    "category": "Electronics",
    "price": 49.99,
    "inventory": 100,
    "inventory_updated_at": "2025-06-04T06:26:00Z",
    "created_at": "2025-06-04T06:26:00Z",
    "updated_at": "2025-06-04T06:26:00Z"
  },
  "source": {
    "version": "2.4.0.Final",
    "connector": "postgresql",
    "name": "oneshop",
    "ts_ms": 1683200760000,
    "snapshot": "true",
    "db": "oneshop",
    "sequence": "[null,\\"23456789\\"]",
    "schema": "public",
    "table": "items",
    "txId": 1234,
    "lsn": 123456789
  },
  "op": "r",
  "ts_ms": 1683200760001
 }
}
```

We can transform this into a more concise format that only contains the essential information we need:

```
{
  "id": 1,
  "name": "Gaming Mouse",
```

CHAPTER 6   REAL-TIME CHANGE DATA CAPTURE WITH KAFKA AND DEBEZIUM

```
  "category": "Electronics",
  "price": 49.99,
  "inventory": 100,
  "inventory_updated_at": "2025-06-04T06:26:00Z"
}
```

To achieve this transformation, we'll create a new connector configuration using Single Message Transforms (SMTs). While we could update the existing connector configuration, this would result in mixed message formats in the topic. For a cleaner implementation when working with OpenSearch later, we'll restart everything from scratch.

To ensure you only see messages with the new format in Kafka:

1. Stop all the containers: docker compose down

2. Start the containers again: docker compose up -d

This gives you a clean slate with only the newly formatted messages in the Kafka topic.

Then create a new connector with the transformation configuration:

```
docker compose exec connect curl -i -X POST <http://
localhost:8083/connectors> \\
  -H "Content-Type: application/json" \\
  -d '{
    "name": "cdc-connector",
    "config": {
      "connector.class": "io.debezium.connector.postgresql.
      PostgresConnector",
      "database.hostname": "postgres",
      "database.port": "5432",
      "database.user": "postgresuser",
      "database.password": "postgrespw",
      "database.dbname": "oneshop",
```

CHAPTER 6   REAL-TIME CHANGE DATA CAPTURE WITH KAFKA AND DEBEZIUM

```
    "database.server.name": "oneshop",
    "table.include.list": "public.items",
    "topic.prefix": "dbz",
    "plugin.name": "pgoutput",
    "slot.name": "cdc_slot",
    "publication.name": "cdc_pub",
    "decimal.handling.mode": "double",
    "include.schema.changes": "false",
    "transforms":"unwrap,extractKey",
    "transforms.unwrap.type":"io.debezium.transforms.
    ExtractNewRecordState",
    "transforms.extractKey.type": "org.apache.kafka.connect.
    transforms.ExtractField$Key",
    "transforms.extractKey.field": "id",
    "key.converter": "org.apache.kafka.connect.json.
    JsonConverter",
    "key.converter.schemas.enable": "false",
    "value.converter": "org.apache.kafka.connect.json.
    JsonConverter",
    "value.converter.schemas.enable": "false"
  }
}'
```

The key additions to the configuration are:

- **transforms**: Specifies two transformation steps: '*unwrap*' and '*extractKey*'

- **transforms.unwrap.type**: Uses Debezium's ExtractNewRecordState transformation to extract just the 'after' state of the record

- **transforms.extractKey.type**: Uses Kafka's ExtractField transformation to use the 'id' field as the message key, which is important for maintaining message order and partitioning. This will be crucial for us in the next section.

- ***.converter.schemas.enable**: Set to "*false*" to remove schema information from both key and value, resulting in cleaner JSON output

- **include.schema.changes**: Set to "*false*" to prevent Debezium from creating additional topics for schema changes, as we only need the data changes

Now, when you check the messages in the topic using either the console consumer or Redpanda Console, you'll see the simplified message format.

Sinking Change Events into OpenSearch

Now that we have our change events flowing into Kafka in a clean format, we can set up the OpenSearch sink connector to consume these events and index them in OpenSearch. This will allow us to maintain a real-time searchable copy of our items data.

OpenSearch is an open-source search and analytics engine that forked from Elasticsearch, providing powerful full-text search capabilities. The OpenSearch Connector for Apache Kafka, developed and distributed by Aiven as open source, comes pre-installed in our Kafka Connect container. This means you don't need to manually download or install the connector – it's ready to use right away.

CHAPTER 6   REAL-TIME CHANGE DATA CAPTURE WITH KAFKA AND DEBEZIUM

First, let's create a connector configuration that maps our Kafka messages to OpenSearch documents:

```
docker compose exec connect curl -i -X POST <http://
localhost:8083/connectors> \\
  -H "Content-Type: application/json" \\
  -d '{
    "name": "opensearch-sink-connector",
    "config": {
      "connector.class": "io.aiven.kafka.connect.opensearch.
      OpensearchSinkConnector",
      "tasks.max": "1",
      "topics": "dbz.public.items",
      "connection.url": "<http://opensearch:9200>",
      "type.name": "_doc",
      "key.ignore": false,
      "schema.ignore": true,
      "name": "opensearch-sink-connector",
      "key.converter": "org.apache.kafka.connect.json.
      JsonConverter",
      "value.converter": "org.apache.kafka.connect.json.
      JsonConverter",
      "key.converter.schemas.enable": "false",
      "value.converter.schemas.enable": "false"
    }
  }'
```

CHAPTER 6   REAL-TIME CHANGE DATA CAPTURE WITH KAFKA AND DEBEZIUM

Let's examine the key configuration parameters:

- **connector.class**: Specifies the OpenSearch sink connector class that handles writing data to OpenSearch

- **topics**: The Kafka topic to consume messages from our Debezium CDC topic

- **opensearch.hosts**: The OpenSearch cluster endpoint, using Docker container name as the hostname.

- **key.ignore**: Set to "false" to preserve message keys when writing to OpenSearch, which helps maintain document IDs

- **schema.ignore**: Set to "true" since we're working with schemaless JSON data

- **key.converter** and **value.converter**: Set to use JSON converter for both keys and values since our Kafka messages are in JSON format

- ***.converter.schemas.enable**: Set to "false" for both key and value converters to handle schemaless JSON data

You should get an output with HTTP status code 201 indicating the success of the operation.

After creating the connector, verify that it's running:

```
docker compose exec connect curl -s localhost:8083/connectors/opensearch-sink-connector/status
```

You should see a JSON response indicating the connector is at the state of RUNNING.

Since we set up our CDC pipeline correctly, all records from the Kafka topic (dbz.public.items) will be automatically indexed into OpenSearch using the **same index name**. Each record's ID field serves as the document

CHAPTER 6   REAL-TIME CHANGE DATA CAPTURE WITH KAFKA AND DEBEZIUM

ID in OpenSearch, ensuring that updates to the same record will modify the corresponding document rather than creating duplicates. To verify that the data is flowing correctly, let's query a single document from OpenSearch using the REST API:

```
docker compose exec connect curl -X GET "<http://opensearch:9200/dbz.public.items/_search?pretty>" \\
  -H 'Content-Type: application/json' \\
  -d '{
    "query": {
      "match_all": {}
    },
    "size": 1
  }'
```

This should return a single document from the `dbz.public.items` index, confirming that our data is successfully flowing from Postgres through Kafka to OpenSearch.

Verify End-to-End CDC Pipeline

Let's verify the end-to-end functionality of our CDC pipeline by making changes in Postgres and observing how they propagate to OpenSearch in real-time.

First, let's connect to Postgres and check the current inventory level. Select the item with id=100.

```
oneshop=# SELECT id, name, inventory FROM items WHERE id = 100;
```

Now, let's check the corresponding document in OpenSearch:

```
docker compose exec connect curl -X GET "<http://opensearch:9200/dbz.public.items/_doc/100?pretty>"
```

```
{
  "_index" : "dbz.public.items",
  "_id" : "100",
  "_version" : 99,
  "_seq_no" : 99,
  "_primary_term" : 1,
  "found" : true,
  "_source" : {
    "updated_at" : 1749029718949442,
    "inventory_updated_at" : 1749029718949442,
    "price" : 263.32,
    "name" : "event disease",
    "created_at" : 1749029718949442,
    "id" : 100,
    "category" : "widgets",
    **"inventory" : 4855**
  }
}
```

The inventory levels should match between Postgres and OpenSearch. Now, let's update the inventory in Postgres:

```
oneshop=# UPDATE items SET inventory = inventory - 1 WHERE id = 100;
```

Finally, verify that the change was captured and propagated to OpenSearch:

```
docker compose exec connect curl -X GET "<http://opensearch:9200/dbz.public.items/_doc/100?pretty>"

{
  "_index" : "dbz.public.items",
  "_id" : "100",
```

```
  "_version" : 1000,
  "_seq_no" : 1000,
  "_primary_term" : 1,
  "found" : true,
  "_source" : {
    "updated_at" : 1749029718949442,
    "inventory_updated_at" : 1749029718949442,
    "price" : 263.32,
    "name" : "event disease",
    "created_at" : 1749029718949442,
    "id" : 100,
    "category" : "widgets",
    **"inventory" : 4854**
  }
}
```

You should see that the inventory level in OpenSearch has been automatically updated to match the new value in Postgres, demonstrating that our CDC pipeline is working correctly.

To stop all containers and clean up resources, run:

```
docker compose down
```

Summary

Congratulations! You have completed your first project using streaming data architecture components like Apache Kafka, Kafka Connect, and Debezium.

In this chapter, we built a real-time CDC pipeline that synchronizes data between Postgres and OpenSearch using Apache Kafka as the messaging backbone. This architecture allows OneShop's data engineering team to maintain a real-time searchable copy of their product inventory data, enabling immediate updates to search results and analytics when changes occur in the source database.

The benefits of this new architecture for OneShop include:

- Instant reflection of inventory changes in the search system
- Reduced load on the primary database by serving search queries from OpenSearch
- Foundation for building real-time analytics and monitoring systems
- Improved scalability and system reliability through event-driven architecture

While we've kept this implementation straightforward for learning purposes, a production environment would need to consider additional factors such as schema evolution, error handling, monitoring, and disaster recovery. You'd also want to implement proper security measures and ensure your CDC pipeline can handle the expected throughput.

In the next chapter, we'll extend this real-time data pipeline to create a dynamic analytics dashboard that provides instant visibility into inventory changes, sales patterns, and other crucial business metrics.

CHAPTER 7

Low-Latency Real-time Analytics Dashboard with ClickHouse

Introduction

OneShop's marketing team plans to run flash sale campaigns effectively by monitoring their performance in real-time. By tracking top-selling items and hourly sales performance, they could adjust inventory levels proactively to prevent stockouts and maintain smooth operation during these time-sensitive events.

In this chapter, we'll explore how to build a real-time analytics pipeline that enables immediate visibility into flash sale campaign performance. We'll extend the Change Data Capture (CDC) pipeline we built in **Chapter 6**, incorporating ClickHouse as our real-time analytics engine and creating an interactive dashboard using Streamlit. This combination will allow OneShop's marketing team to monitor and optimize their flash sale campaigns as they happen.

CHAPTER 7   LOW-LATENCY REAL-TIME ANALYTICS DASHBOARD WITH CLICKHOUSE

Flash Sale Performance Dashboard

The sales dashboard we built in **Chapter 4** with Apache Superset provided only periodic insights into sales performance. The metrics shown in the dashboard could be anywhere from 5 minutes to 24 hours old. While this is acceptable for daily sales performance reports, we need a dashboard that displays flash sale metrics in real time, giving the marketing team accurate insights to adjust campaign parameters on the fly.

To achieve this goal, we will extend the CDC pipeline we built in **Chapter 6** with two downstream components:

- Configure Clickhouse database to ingest the `purchases` table's change data feed from Kafka and maintain an incrementally updated materialized view of flash sale performance.

- Build a real-time analytics dashboard with Streamlit to connect to Clickhouse and visualize the materialized view.

Before we get hands-on, let's understand what real-time analytics is and how it differs from the traditional batch analytics we covered in Part 1.

What Is Real-Time Analytics?

Real-time analytics refers to the ability to collect, process, and analyze data as soon as it becomes available, typically within seconds or milliseconds of the events occurring. Unlike batch analytics, which processes data in scheduled intervals, real-time analytics provides immediate insights that enable organizations to respond quickly to changing conditions.

Business Value Over Batch Analytics

The transition from batch to real-time analytics represents a fundamental shift in how organizations derive value from their data. While batch processing remains valuable for historical analysis and complex computations, real-time analytics enables businesses to capitalize on immediate opportunities and respond to threats as they emerge. This capability has become increasingly critical in today's fast-paced digital economy.

Real-time analytics provides several advantages over traditional batch processing:

- **Immediate decision making:** Enables organizations to respond to events and opportunities as they happen
- **Proactive problem resolution:** Helps identify and address issues before they escalate
- **Enhanced customer experience:** Allows for personalized, context-aware interactions
- **Operational efficiency:** Reduces lag between data generation and action-taking

The value of real-time analytics continues to grow as businesses face increasing pressure to make faster, data-driven decisions in response to rapidly changing market conditions and customer expectations.

Key Components of Real-time Analytics Systems

A real-time analytics system comprises several essential components working together to deliver instant insights. These components must be carefully designed and integrated to handle high-throughput data streams while maintaining low latency and data accuracy. Here are the fundamental building blocks:

- **Data ingestion layer:** Handles continuous streams of data from various sources (e.g., CDC feeds, application events, IoT sensors)
- **Stream processing engine:** Processes incoming data streams in real-time, applying transformations and aggregations
- **Storage layer:** Specialized databases optimized for real-time querying and analytics
- **Visualization layer:** Dashboards and tools that can update metrics in real-time

{ a visual goes here}

Now that we understand the fundamentals of real-time analytics and its importance in modern data architectures, let's dive into implementing our flash sale analytics dashboard. We'll use Clickhouse as our real-time analytics database due to its excellent performance characteristics for real-time queries and its ability to handle high-throughput data ingestion.

Let's begin with the initial setup and configuration of our real-time analytics infrastructure.

Before You Begin

You will find the code for this chapter located in the chapter-07 folder. If you haven't set everything up yet, refer to the **Prerequisites** section in **Chapter 1** for more information.

Navigate to the project folder on a terminal:

```
cd <repository_root>/chapter-07
```

In this folder, you will find everything you need to deploy a minimal real-time analytics dashboard on Docker Compose.

CHAPTER 7 LOW-LATENCY REAL-TIME ANALYTICS DASHBOARD WITH CLICKHOUSE

Breakdown of the Docker-Compose File

The chapter-07/docker-compose.yml file defines the services that make up our analytics dashboard.

```
services:
  postgres:
    image: debezium/postgres:17
    hostname: postgres
    container_name: postgres
    ports:
      - 5432:5432
    environment:
      - POSTGRES_USER=postgresuser
      - POSTGRES_PASSWORD=postgrespw
      - POSTGRES_DB=oneshop
      - PGPASSWORD=postgrespw
    volumes:
      - ./postgres/postgres_bootstrap.sql:/docker-entrypoint-
        initdb.d/postgres_bootstrap.sql
    healthcheck:
      test: ["CMD-SHELL", "pg_isready -U postgresuser -d
      oneshop"]
      interval: 1s
      start_period: 60s

  zookeeper:
    container_name: zookeeper
    image: bitnami/zookeeper:3.8
    environment:
      ALLOW_ANONYMOUS_LOGIN: "yes"
    ports:
      - "2181:2181"
```

CHAPTER 7   LOW-LATENCY REAL-TIME ANALYTICS DASHBOARD WITH CLICKHOUSE

```yaml
  kafka:
    container_name: kafka
    image: bitnami/kafka:3.6
    depends_on:
      - zookeeper
    ports:
      - "9092:9092"
    environment:
      KAFKA_CFG_ZOOKEEPER_CONNECT: zookeeper:2181
      KAFKA_CFG_LISTENERS: PLAINTEXT://:9092
      KAFKA_CFG_ADVERTISED_LISTENERS: PLAINTEXT://kafka:9092
      KAFKA_CFG_AUTO_CREATE_TOPICS_ENABLE: "true"
      ALLOW_PLAINTEXT_LISTENER: "yes"

  connect:
    container_name: connect
    image: debezium/connect:2.5
    depends_on:
      - kafka
      - postgres
    ports:
      - "8083:8083"
    environment:
      BOOTSTRAP_SERVERS: kafka:9092
      GROUP_ID: 1
      CONFIG_STORAGE_TOPIC: connect-configs
      OFFSET_STORAGE_TOPIC: connect-offsets
      STATUS_STORAGE_TOPIC: connect-status
      KEY_CONVERTER: org.apache.kafka.connect.json.JsonConverter
      VALUE_CONVERTER: org.apache.kafka.connect.json.JsonConverter
```

```
      INTERNAL_KEY_CONVERTER: org.apache.kafka.connect.json.
      JsonConverter
      INTERNAL_VALUE_CONVERTER: org.apache.kafka.connect.json.
      JsonConverter
      CONNECT_REST_ADVERTISED_HOST_NAME: connect
      PLUGIN_PATH: /kafka/connect
  redpanda-console:
    image: docker.redpanda.com/redpandadata/console:latest
    container_name: redpanda-console
    ports:
      - "8080:8080"
    environment:
      KAFKA_BROKERS: kafka:9092
  clickhouse:
    image: clickhouse/clickhouse-server
    container_name: clickhouse
    volumes:
      - ./clickhouse/default-password.xml:/etc/clickhouse-
        server/users.d/default-password.xml
    ports:
      - "9000:9000"
      - "8123:8123"
    ulimits:
      nproc: 65535
      nofile:
        soft: 262144
        hard: 262144
  flashsale-loadgen:
    build: flashsale-loadgen
    container_name: flashsale-loadgen
```

```
      init: true
      depends_on:
        postgres: {condition: service_healthy}
    streamlit:
      build: streamlit
      container_name: streamlit
      init: true
      ports:
        - "8501:8501"
      depends_on:
        - clickhouse
```

The file includes several services we previously defined in **Chapter 6**, along with some new additions. Let's examine the new services a little further:

- `clickhouse` – A single-node Clickhouse server that serves as our real-time analytics database.
- `streamlit` – The real-time analytics dashboard that visualizes flash sale metrics.
- `flashsale-loadgen` – A Python script to populate the purchases Postgres table.

Running the Setup

From the root level of the `chapter-07` directory, start all containers by running:

```
docker-compose up -d --build
```

The above command will build the flashsale-loadgen container first, and then start other containers in the background. Although this Docker Compose setup isn't designed for large-scale production use, it offers a convenient local environment to explore all the key components of a real-time analytics system in one place.

Source Database and Schema

We'll use the same oneshop schema from **Chapter 6**, with one modification: we're adding a new field called campaign_id to the purchases table to track which campaign led to each conversion. For this chapter, we'll focus only on the purchases table since it records the sales.

You'll find the modified database initialization script in the ./chapter-07/postgres folder, which creates the following schema for the purchases table.

```
CREATE TABLE IF NOT EXISTS purchases
(
    id SERIAL PRIMARY KEY,
    user_id BIGINT REFERENCES users(id),
    item_id BIGINT REFERENCES items(id),
    campaign_id ,
    status SMALLINT DEFAULT 1,
    quantity INT DEFAULT 1,
    purchase_price DECIMAL(12,2),
    deleted BOOLEAN DEFAULT FALSE,
    created_at TIMESTAMP DEFAULT CURRENT_TIMESTAMP,
    updated_at TIMESTAMP DEFAULT CURRENT_TIMESTAMP
);
```

Additionally, the loadgen service has been modified to generate purchase events with FLASH2025 as the campaign code. We'll use this code later to filter purchases specific to this campaign.

Creating the Change Data Feed for the purchases Table

Just as we did in **Chapter 6** for the items table, let's create a new change feed for the purchases table in the oneshop database.

Create a Debezium source connector for the purchases table by running:

```
docker compose exec connect curl -i -X POST <http://localhost:8083/connectors> \\
  -H "Content-Type: application/json" \\
  -d '{
    "name": "cdc-connector",
    "config": {
      "connector.class": "io.debezium.connector.postgresql.PostgresConnector",
      "database.hostname": "postgres",
      "database.port": "5432",
      "database.user": "postgresuser",
      "database.password": "postgrespw",
      "database.dbname": "oneshop",
      "database.server.name": "oneshop",
      "table.include.list": "public.purchases",
      "topic.prefix": "dbz",
      "plugin.name": "pgoutput",
      "slot.name": "cdc_slot",
      "publication.name": "cdc_pub",
```

CHAPTER 7   LOW-LATENCY REAL-TIME ANALYTICS DASHBOARD WITH CLICKHOUSE

```
    "decimal.handling.mode": "double",
    "include.schema.changes": "false",
    "transforms":"unwrap",
    "transforms.unwrap.type":"io.debezium.transforms.
    ExtractNewRecordState",
    "key.converter": "org.apache.kafka.connect.json.
    JsonConverter",
    "key.converter.schemas.enable": "false",
    "value.converter": "org.apache.kafka.connect.json.
    JsonConverter",
    "value.converter.schemas.enable": "false"
  }
}'
```

The connector configuration is similar to what we used in **Chapter 6**, with one difference: the table.include.list now specifies the purchases table as the source. Since we've covered these settings before, we won't explain them again here.

Next, check whether a new Kafka topic has been created to receive the change feed from the purchases table.

```
docker compose exec kafka kafka-topics.sh --bootstrap-server localhost:9092 --list
```

You should see the new topic dbz.public.purchases in the output. If you don't, check the Kafka Connect logs by running:

```
docker compose logs -f connect
```

Let's check if the change events are being published to the dbz.public.purchases topic. You can use either of these two methods:

175

Using the Kafka Console Consumer

```
docker compose exec kafka kafka-console-consumer.sh \\
  --bootstrap-server localhost:9092 \\
  --topic dbz.public.purchases \\
  --from-beginning
```

Using the Redpanda Console

Open http://localhost:8080 in your browser to access the Redpanda Console. Navigate to the Topics section and click on `dbz.public.purchases` to view the messages in real-time.

You should see a continuous stream of JSON messages representing purchase events being generated by the `flashsale-loaden` service.

A quick peek at a single change event from the `dbz.public.purchases` topic will give us this format:

```
{
    "id": 51,
    "user_id": 411,
    "item_id": 300,
    "campaign_id": "FLASH2025",
    "status": 1,
    "quantity": 1,
    "purchase_price": 49.99,
    "deleted": false,
    "created_at": 1748502285528056,
    "updated_at": 1748502285528056
}
```

CHAPTER 7   LOW-LATENCY REAL-TIME ANALYTICS DASHBOARD WITH CLICKHOUSE

Configuring Clickhouse

Now that we have our change data feed set up, let's configure Clickhouse to store and process this data efficiently. We'll set up tables and materialized views that will enable us to analyze flash sale performance metrics in real-time.

ClickHouse is an open-source column-oriented database management system that excels at real-time analytics processing. It's designed for high-performance analytics workloads and can handle billions of rows with sub-second query response times. Its columnar storage format and parallel processing capabilities are well-suited for our flash sale analytics use case.

The Kafka table engine enables ClickHouse to read directly from a Kafka topic. Let's configure a Kafka table engine to ingest purchase change events into ClickHouse.

We can do this by logging into the ClickHouse Client – a command-line interface (CLI) that lets users query tables and perform administrative tasks.

```
docker compose exec clickhouse clickhouse-client
```

When prompted, enter the password "mysecret."

We've configured the ClickHouse server with a password for the `default` user by mounting the `./clickhouse/default-password.xml` file into the ClickHouse container:

```
volumes:
    - ./clickhouse/default-password.xml:/etc/clickhouse-
      server/users.d/default-password.xml
```

Here's how the password is set in the configuration file:

```
<clickhouse>
    <users>
        <default>
```

```xml
            <password>mysecret</password>
        </default>
    </users>
</clickhouse>
```

After logging in, we'll define the source table using the Kafka engine by running the following queries:

```sql
CREATE DATABASE oneshop;
USE oneshop;

CREATE TABLE purchases_raw (
    id              Int32,
    user_id         Int64,
    item_id         Int64,
    campaign_id     String,
    status          Int16,
    quantity        Int32,
    purchase_price  Decimal(12, 2),
    deleted         Bool,
    created_at      DateTime64(6),
    updated_at      DateTime64(6)
)
ENGINE = Kafka
SETTINGS
    kafka_broker_list = 'kafka:9092',
    kafka_topic_list = 'dbz.public.purchases',
    kafka_group_name = 'clickhouse-consumer-group',
    kafka_format = 'JSONEachRow',
    kafka_num_consumers = 1;
```

The `purchases_raw` table stores raw purchase events from the `dbz.public.purchases` topic. The table's SETTINGS configuration specifies that ClickHouse should use the Kafka engine for data ingestion.

The `kafka_format = 'JSONEachRow'` setting tells ClickHouse to expect and parse JSON-formatted messages from Kafka. Since both Kafka and ClickHouse services are running in the same Docker network, the connectivity between them is seamless using the internal DNS name `kafka:9092`.

We can't directly query the `purchases_raw` table as it is a Kafka engine table. These tables in ClickHouse only handle streaming data ingestion and maintain a local offset value that advances with each read operation. After data is read and the offset advances, you cannot access that data again through subsequent queries unless you manually reset the offsets – a practice that's typically discouraged in production systems.

We need another table, `purchases`, to permanently store the purchase data on disk using ClickHouse's MergeTree() storage engine. MergeTree is ClickHouse's most versatile and powerful table engine for analytical workloads. It provides efficient data storage and retrieval by sorting data based on the primary key. The engine automatically merges data parts in the background to optimize storage and query performance.

Next, we create a materialized view, `purchases_mv`, to ensure all data flowing through the `purchases_raw` Kafka table is automatically persisted to our MergeTree-based purchases table. This gives us both real-time access to the latest data and the ability to perform historical analysis efficiently.

Run the following query to define both the table and the materialized view.

```
CREATE TABLE purchases
(
    id              Int32,
    user_id         Int64,
    item_id         Int64,
    campaign_id     String,
    status          Int16,
```

CHAPTER 7 LOW-LATENCY REAL-TIME ANALYTICS DASHBOARD WITH CLICKHOUSE

```
    quantity         Int32,
    purchase_price   Decimal(12, 2),
    deleted          Bool,
    created_at       DateTime64(6),
    updated_at       DateTime64(6)
)
ENGINE = MergeTree
ORDER BY (id);

-- Create a materialized view to populate the MergeTree table
CREATE MATERIALIZED VIEW mv_purchases
TO purchases
AS
SELECT * FROM purchases_raw;
```

All user-facing queries from dashboards or applications will interact with the purchases table from this point forward.

To verify whether the materialized view is getting populated, run the following query in the playground.

```
SELECT * FROM purchases;
```

It should print the first 100 rows in the purchases table.

Let's test two analytical queries that will power our dashboard.

The first query shows the top ten selling items, helping us identify the best-performing products during the flash sale campaign by units sold and revenue generated.

```
SELECT
    item_id,
    SUM(quantity) AS total_quantity,
    SUM(quantity * purchase_price) AS total_revenue
```

```
FROM purchases
GROUP BY item_id
ORDER BY total_revenue DESC
LIMIT 10
```

The second query provides an hourly breakdown of sales, which will generate a bar chart in the next section.

```
SELECT
    toStartOfHour(created_at) AS hour,
    SUM(quantity) AS total_sales
FROM purchases
WHERE deleted = 0
  AND created_at >= now() - INTERVAL 24 HOUR
GROUP BY hour
ORDER BY hour ASC
```

Running the Streamlit Dashboard

We've now reached the final stage of our real-time analytics pipeline, where we'll visualize the purchase metrics stored in ClickHouse using a Streamlit dashboard that updates in real-time. While ClickHouse offers integration with various BI tools like Tableau, Power BI, or Grafana, we've specifically chosen Streamlit for this implementation.

Streamlit is a Python-based framework that enables data engineers to quickly build interactive data applications with minimal code. Its ability to handle real-time data updates, combined with its straightforward API and built-in caching mechanisms, makes it an excellent choice for rapidly prototyping and deploying data-driven dashboards. Plus, its deep integration with popular data science libraries like Pandas and Plotly allows for sophisticated data manipulation and visualization capabilities.

CHAPTER 7 LOW-LATENCY REAL-TIME ANALYTICS DASHBOARD WITH CLICKHOUSE

The Streamlit framework can be installed locally simply by running `pip install streamlit` command. To make things easier, we've already installed it in the Streamlit container. When the container starts, the dashboard is also deployed with Streamlit. Check the chapter-07/streamlit/Dockerfile to see the installation instructions and dependencies (pandas and plotly) required for the dashboard.

You can find the completed dashboard code inside the chapter-07/streamlit/app.py.

```python
import streamlit as st
import pandas as pd
import plotly.express as px
from clickhouse_connect import get_client
from datetime import datetime, timedelta

# Set up ClickHouse connection
CLICKHOUSE_HOST = 'clickhouse'    # Replace with your
                                  ClickHouse host
CLICKHOUSE_PORT = 8123            # Default HTTP port
CLICKHOUSE_USER = 'default'
CLICKHOUSE_PASSWORD = 'mysecret'

client = get_client(host=CLICKHOUSE_HOST, port=CLICKHOUSE_PORT,
username=CLICKHOUSE_USER, password=CLICKHOUSE_PASSWORD)

st.set_page_config(page_title="Flash Sale Performance
Dashboard", layout="wide")
st.title("🛒 Flash Sale [FLASH2025] Performance")

# --- 1. Hourly breakdown of sales volume for the last 24
hours ---

st.header("Hourly Sales Volume (Last 24 Hours)")
```

```python
hourly_sales_query = """
SELECT
    toStartOfHour(created_at) AS hour,
    SUM(quantity) AS total_sales
FROM oneshop.purchases
WHERE deleted = 0
  AND created_at >= now() - INTERVAL 24 HOUR
GROUP BY hour
ORDER BY hour ASC
"""

hourly_sales_df = client.query_df(hourly_sales_query)

fig = px.bar(hourly_sales_df, x="hour", y="total_sales",
title="Hourly Sales Volume (Past 24 Hours)", labels={"hour":
"Hour", "total_sales": "Quantity Sold"})
st.plotly_chart(fig, use_container_width=True)

# --- 2. Top 10 most selling product IDs with their revenue ---

st.header("Top 10 Selling Product IDs by Revenue")
top_products_query = """
SELECT
    item_id,
    SUM(quantity) AS total_quantity,
    SUM(quantity * purchase_price) AS total_revenue
FROM oneshop.purchases
GROUP BY item_id
ORDER BY total_revenue DESC
LIMIT 10
"""

top_products_df = client.query_df(top_products_query)
st.dataframe(top_products_df, use_container_width=True)
```

CHAPTER 7 LOW-LATENCY REAL-TIME ANALYTICS DASHBOARD WITH CLICKHOUSE

Let's break down the Python code and understand what's happening:

1. First, the code establishes a connection to ClickHouse using the clickhouse_connect driver. It sets up connection parameters like host, port, username, and password to connect to our ClickHouse instance running in Docker.

2. For the bar chart visualization, the code queries ClickHouse for hourly sales data from the last 24 hours. It uses Plotly Express (px.bar) to create an interactive bar chart showing sales volume over time. The chart is then rendered in the Streamlit interface using st.plotly_chart().

3. Finally, the code generates a table of top-selling items by executing a query that aggregates purchase data by item_id, calculating total quantity and revenue. The results are displayed in a Streamlit dataframe (st.dataframe()) showing the top 10 products by revenue.

Finally, you can access the dashboard by visiting http://localhost:8501/. You should see something similar to this.

CHAPTER 7   LOW-LATENCY REAL-TIME ANALYTICS DASHBOARD WITH CLICKHOUSE

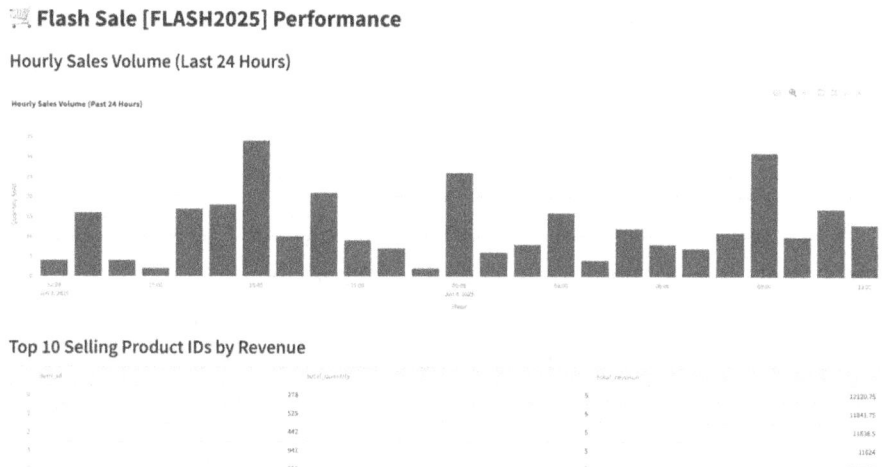

Figure 7-1. Flash sale performance dashboard

Summary

Congratulations! You've completed the second project focused on streaming data architecture. You now have a functioning real-time analytics dashboard with a CDC pipeline.

In this chapter, we explored building a real-time analytics dashboard to monitor flash sale performance. We set up a Docker Compose environment with essential services: Postgres, Kafka, Kafka Connect with Debezium, ClickHouse, and Streamlit. Using Debezium, we configured change data capture from the purchases table to stream events to Kafka in real time.

For data processing and storage, we implemented ClickHouse with raw tables for Kafka stream ingestion and materialized views for efficient querying. We then created a Streamlit dashboard that visualizes flash sale metrics in real time, demonstrating how modern data tools can create powerful analytics solutions.

CHAPTER 7   LOW-LATENCY REAL-TIME ANALYTICS DASHBOARD WITH CLICKHOUSE

With this real-time analytics dashboard in place, both the marketing team and OneShop supply chain managers can monitor sales performance instantly, adjust stock levels dynamically, and prevent stockouts during high-demand periods – all without the delays typical of traditional batch processing systems.

The seamless integration between marketing analytics and supply chain operations enables more coordinated campaign execution. Teams can make data-driven decisions about inventory allocation, reorder timing, and distribution center management within minutes instead of hours or days, significantly improving operational efficiency during time-sensitive sales events.

CHAPTER 8

Streaming ETL and Anomaly Detection with Apache Flink

Introduction

OneShop's IT department frequently receives support tickets regarding suspicious login attempts from customers who believe their account credentials have been compromised and used for unauthorized purchases. To address this security concern, the IT department has requested the data engineering team's assistance in building a login anomaly detection system.

In this chapter, we'll develop a real-time login anomaly detection system using Apache Flink, a distributed stream processing framework. We'll use Apache Kafka to collect login events from web and mobile devices, process them through a Flink cluster, and create detection jobs to identify suspicious activities.

By the end of this chapter, you'll learn how to enhance streaming data pipelines through integration with Flink for real-time analysis and sub-second level responsiveness.

CHAPTER 8   STREAMING ETL AND ANOMALY DETECTION WITH APACHE FLINK

Login Anomaly Detection System

Building on their successful implementation of streaming data architectures at OneShop, the data engineering team has developed extensive expertise in real-time data pipelines, change data capture (CDC), and real-time analytics dashboards. Their new project focuses on creating a low-latency anomaly detection system using Apache Flink.

The pipeline first collects login events from web and mobile applications through Kafka. Apache Flink then transforms and enriches these events before a specialized job analyzes them for unusual login activities and triggers appropriate alerting workflows.

Before we dive in, let's explore what stateful stream processing is and how Apache Flink implements it.

Stateful Stream Processing and Apache Flink

Stateful stream processing is a paradigm where processing tasks maintain and use state information across multiple events over time. Unlike stateless processing, which treats each event independently, stateful processing remembers previous events and their context to make more informed decisions. This capability enables complex operations like windowing, aggregations, and pattern detection in streaming data.

Stateful processing offers several powerful capabilities that make it ideal for real-time data analysis. These include the ability to detect patterns and trends over time, support complex event processing and real-time analytics, enhance accuracy in anomaly detection and fraud prevention, and maintain user sessions and context. These features enable systems to make intelligent decisions based on historical data and current context.

CHAPTER 8   STREAMING ETL AND ANOMALY DETECTION WITH APACHE FLINK

Despite its advantages, implementing stateful processing comes with significant technical challenges that need to be carefully addressed. Organizations must manage state consistency across distributed systems, implement robust state recovery mechanisms for handling failures, ensure their storage solutions can scale with growing data volumes, and maintain exactly-once processing semantics to guarantee data accuracy. These challenges require careful architectural planning and implementation.

Apache Flink emerged as a solution to these challenges. Originally developed at the Technical University of Berlin in 2010 as a research project called Stratosphere, Flink became an Apache top-level project in 2014. It was designed specifically for stateful computations over data streams.

Flink's core concepts include:

- **Streams:** Unbounded or bounded sequences of data records

- **State:** Local variables and data structures maintained by operators

- **Time:** Support for event time, processing time, and ingestion time semantics

- **Transformations:** Operations that convert one or more streams into new streams

What sets Flink apart is its ability to handle both stream and batch processing with the same execution engine, treating batch processing as a special case of stream processing with bounded data sets. This unified approach, combined with its robust state management capabilities and seamless integration with streaming data sources like Apache Kafka, makes it particularly suitable for building real-time applications like OneShop's login anomaly detection system.

Having covered the fundamentals of stream processing, we'll now implement our anomaly detection system, starting with the setup and configuration of our real-time data infrastructure.

CHAPTER 8   STREAMING ETL AND ANOMALY DETECTION WITH APACHE FLINK

Before You Begin

You will find the code for this chapter located in the chapter-08 folder. If you haven't set everything up yet, refer to the **Prerequisites** section in **Chapter 1** for more information.

Navigate to the project folder on a terminal:

```
cd <repository_root>/chapter-08
```

In this folder, you will find everything you need to deploy a minimal Apache Kafka and Flink cluster with other supplementary tools.

Breakdown of the Docker-Compose File

The chapter-08/docker-compose.yml file defines the services that make up our anomaly detection system.

```yaml
services:
  zookeeper:
    container_name: zookeeper
    image: bitnami/zookeeper:3.8
    environment:
      ALLOW_ANONYMOUS_LOGIN: "yes"
    ports:
      - "2181:2181"

  kafka:
    container_name: kafka
    image: bitnami/kafka:3.6
    depends_on:
      - zookeeper
    ports:
      - "9092:9092"
```

```yaml
  environment:
    KAFKA_CFG_ZOOKEEPER_CONNECT: zookeeper:2181
    KAFKA_CFG_LISTENERS: PLAINTEXT://:9092
    KAFKA_CFG_ADVERTISED_LISTENERS: PLAINTEXT://kafka:9092
    KAFKA_CFG_AUTO_CREATE_TOPICS_ENABLE: "true"
    ALLOW_PLAINTEXT_LISTENER: "yes"
jobmanager:
  image: flink:1.16.0-scala_2.12-java11
  container_name: jobmanager
  ports:
    - 8081:8081
  command: jobmanager
  environment:
    - |
      FLINK_PROPERTIES=
      jobmanager.rpc.address: jobmanager
taskmanager:
  image: flink:1.16.0-scala_2.12-java11
  container_name: taskmanager
  depends_on:
    - jobmanager
  command: taskmanager
  scale: 1
  environment:
    - |
      FLINK_PROPERTIES=
      jobmanager.rpc.address: jobmanager
      taskmanager.numberOfTaskSlots: 20
```

CHAPTER 8   STREAMING ETL AND ANOMALY DETECTION WITH APACHE FLINK

Let's examine these services a little further:

- **Zookeeper:** Manages and coordinates the Kafka cluster, handling configuration information, naming, and synchronization between Kafka brokers.

- **Kafka:** A Message broker that handles the streaming data pipeline, collecting login events from various sources and making them available for processing.

- **Jobmanager:** Flink's central coordinator that manages the execution of streaming jobs, handles job scheduling, and coordinates checkpoints for fault tolerance.

- **Task Manager:** Worker node in the Flink cluster that executes the actual data processing tasks. Configured with 20 task slots to handle parallel processing of streaming data.

Running the Setup

From the root level of the `chapter-08` directory, start all containers by running:

`docker-compose up -d --build`

The above command will build `loadgen` container first, and then start other containers in the background.

Creating Kafka Topics

We need to create three Kafka topics to handle different stages of our login event processing:

- `login-events`: The source topic that receives raw login events from web and mobile applications
- `login-events-enriched`: Topic for storing enriched login information, including country and the device type
- `login-anomalies`: Destination topic for detected suspicious login attempts

To create these topics, execute the following commands:

```
docker-compose exec kafka kafka-topics.sh --create \\
    --bootstrap-server kafka:9092 \\
    --topic login-events \\
    --partitions 1 \\
    --replication-factor 1

docker-compose exec kafka kafka-topics.sh --create \\
    --bootstrap-server kafka:9092 \\
    --topic login-events-enriched \\
    --partitions 1 \\
    --replication-factor 1

docker-compose exec kafka kafka-topics.sh --create \\
    --bootstrap-server kafka:9092 \\
    --topic login-anomalies \\
    --partitions 1 \\
    --replication-factor 1
```

CHAPTER 8   STREAMING ETL AND ANOMALY DETECTION WITH APACHE FLINK

We're using 1 partition for each topic to keep things simple in this development setup, with a replication factor of 1 since we're running a single-broker environment. In a production environment, you'd want multiple partitions and a higher replication factor for fault tolerance.

Creating Flink SQL Tables

Next, we are going to create some tables in Flink. We will use Flink SQL for that.

Flink SQL is a powerful declarative language that simplifies stream processing by allowing developers to express complex data transformations and analytics using familiar SQL syntax. Rather than writing detailed procedural code to process data streams, developers can leverage their existing SQL knowledge to define what they want to achieve, letting Flink handle the underlying implementation details.

The declarative nature of Flink SQL significantly enhances developer productivity by abstracting away the complexities of distributed stream processing. Developers can focus on business logic and data transformations instead of worrying about low-level stream processing concepts, state management, or parallel execution. This abstraction not only speeds up development but also reduces the likelihood of errors that might occur when writing complex stream processing logic manually.

The complete SQL script can be found in the `./chapter-08/flink-sql/init.sql` file. Let's examine each table definition individually to understand the structure.

The first table is `login_events`, which ingests raw login events from the Kafka source topic, `login-events`.

```
CREATE TABLE login_events (
  user_id STRING,
  `timestamp` TIMESTAMP_LTZ(3),
  ip STRING,
```

```
  device STRING,
  platform STRING,
  user_agent STRING,
  WATERMARK FOR `timestamp` AS `timestamp` - INTERVAL
  '5' SECOND
) WITH (
  'connector' = 'kafka',
  'topic' = 'login-events',
  'properties.bootstrap.servers' = 'kafka:9092',
  'format' = 'json',
  'scan.startup.mode' = 'earliest-offset'
);
```

This table uses JSON format for data serialization and includes a watermark definition for handling event time processing with a 5-second allowed lateness. It's set up to read from the earliest offset in the Kafka topic, ensuring no messages are missed when the system starts.

Second, we have the `login-events-enriched` table, which ingests the enriched login events from the `login-events-enriched` topic.

```
CREATE TABLE login_events_enriched (
  user_id STRING,
  `timestamp` TIMESTAMP_LTZ(3),
  ip STRING,
  platform STRING,
  device_type STRING,
  country STRING,
  city STRING
) WITH (
  'connector' = 'kafka',
  'topic' = 'login-events-enriched',
```

```
  'properties.bootstrap.servers' = 'kafka:9092',
  'format' = 'json'
);
```

The configuration is pretty much similar to the first one. Notice the enriched fields, including device_type and country.

Then we define a static table for GeoIP lookups. It maps IP prefixes to the corresponding country/city, which is loaded from a CSV file inside the Flink container.

```
CREATE TABLE ip_geo (
  ip_prefix STRING,
  country STRING,
  city STRING
) WITH (
  'connector' = 'filesystem',
  'path' = 'file:///opt/flink/ip_geo.csv',
  'format' = 'csv'
);
```

The contents inside the ip_geo.csv look something like this:

```
192.168.1.,UK,London
203.0.113.,US,New York
10.0.0.,DE,Berlin
```

Our last table is the login_anomalies table, where the detected login anomalies are written.

```
CREATE TABLE login_anomalies (
  user_id STRING,
  `timestamp` TIMESTAMP_LTZ(3),
  reason STRING,
  country STRING
```

```
) WITH (
  'connector' = 'kafka',
  'topic' = 'login-anomalies',
  'properties.bootstrap.servers' = 'kafka:9092',
  'format' = 'json'
);
```

Then we write two SQL queries that do the enrichment and anomaly detection. They are in the INSERT INTO <table1> SELECT FROM <table2> format.

```
INSERT INTO login_events_enriched
SELECT
  user_id,
  `timestamp`,
  ip,
  platform,
  CASE
    WHEN LOWER(device) LIKE '%iphone%' OR LOWER(device) LIKE
    '%android%' THEN 'mobile'
    ELSE 'desktop'
  END AS device_type,
  g.country,
  g.city
FROM login_events
LEFT JOIN ip_geo AS g ON ip LIKE CONCAT(g.ip_prefix, '%');
```

This query enriches the login events with device type and geographical information. The device type is determined using a CASE statement that checks if the device string contains "iphone" or "android" (case-insensitive) to classify it as either "mobile" or "desktop." The geographical enrichment is done through a LEFT JOIN with the ip_geo table, where it matches the IP address prefix pattern to add the corresponding country and city information.

CHAPTER 8   STREAMING ETL AND ANOMALY DETECTION WITH APACHE FLINK

The final query in the script does the actual anomaly detection.

```
-- Anomaly detection: new country in past 7 days
INSERT INTO login_anomalies
SELECT
  e.user_id,
  e.timestamp,
  'Login from new country' AS reason,
  e.country
FROM login_events_enriched e
LEFT JOIN (
  SELECT DISTINCT user_id, country
  FROM login_events_enriched
  WHERE `timestamp` BETWEEN CURRENT_TIMESTAMP - INTERVAL '7'
  DAY AND CURRENT_TIMESTAMP
) AS recent
ON e.user_id = recent.user_id AND e.country = recent.country
WHERE recent.country IS NULL;
```

This query detects anomalies by identifying logins from countries that a user hasn't logged in from in the past 7 days. Let's break down how it works:

- The subquery creates a list of all distinct user-country combinations from the past 7 days using the window:

    `CURRENT_TIMESTAMP - INTERVAL '7' DAY AND CURRENT_TIMESTAMP`

- The main query then LEFT JOINs the current login event with this historical data, matching on both user_id and country

- If `recent.country IS NULL`, it means there's no record of the user logging in from this country in the past week, indicating a potential security risk

CHAPTER 8   STREAMING ETL AND ANOMALY DETECTION WITH APACHE FLINK

- When such cases are found, it generates an anomaly record with

  - The user's ID

  - The timestamp of the suspicious login

  - The reason "Login from a new country"

  - The suspicious country location

This approach helps identify potential account compromises where attackers might be accessing accounts from different geographical locations than the legitimate users' usual login locations.

To execute the SQL queries we defined above, we'll use the Flink SQL client.

First, launch the Flink SQL shell with this command:

```
docker-compose exec jobmanager ./bin/sql-client.sh
```

You will see a CLI with a giant squirrel logo displayed.

As illustrated in Figure 8-1, the Flink anomaly detection topology processes streaming login events in real time.

CHAPTER 8   STREAMING ETL AND ANOMALY DETECTION WITH APACHE FLINK

Figure 8-1. *Flink CLI with a giant squirrel logo*

First, copy the content from ./flink-sql/create-tables.sql and run each statement block individually. The Flink SQL shell cannot execute multiple SQL statements simultaneously. These statements create Flink tables. To verify they were created successfully, run:

SHOW TABLES;

Next, execute the SQL statements in ./flink-sql/insert-jobs.sql to create continuous data pipelines for the login_events_enriched and login_anomalies tables. After running these INSERT statements, you'll see them operating as persistent streaming jobs.

You can verify the jobs are running by visiting the Flink Web UI at http://localhost:8081. You should see two running jobs – one for enrichment and one for anomaly detection.

Simulating Login Events and Verification

Now that our anomaly detection jobs are deployed and running in Flink, let's produce a stream of login events using a Python script to simulate live user logins from web and mobile devices.

You can find the simulator script inside ./chapter-08/simulator/generate_load.py file.

```
from kafka import KafkaProducer
import json
from datetime import datetime
import time
import random

producer = KafkaProducer(bootstrap_servers='localhost:9092',
                        value_serializer=lambda x: json.
                        dumps(x).encode('utf-8'))

devices = ['iPhone 15', 'Samsung Galaxy', 'MacBook Pro', 'Windows Laptop']
ips = ['192.168.1.100', '203.0.113.22', '10.0.0.99']
platforms = ['ios', 'android', 'web']
```

CHAPTER 8   STREAMING ETL AND ANOMALY DETECTION WITH APACHE FLINK

```
while True:
    event = {
        "user_id": f"user_{random.randint(1,5)}",
        "timestamp": datetime.utcnow().isoformat() + "Z",
        "ip": random.choice(ips),
        "device": random.choice(devices),
        "platform": random.choice(platforms),
        "user_agent": "Mozilla/5.0"
    }
    producer.send('login-events', value=event)
    print("Sent:", event)
    time.sleep(2)
```

Let's break down code.

First, the script sets up the necessary imports and configurations. It uses the kafka-python library to interact with Kafka, along with other standard modules for handling JSON, datetime operations, and random generation. A Kafka producer is initialized to connect to localhost:9092, with JSON serialization enabled for the message values.

The script then defines sample data configurations through several lists. These include different types of devices (like iPhone 15 and MacBook Pro), sample IP addresses representing different locations, and various platforms (ios, android, web). These lists provide the pool of values from which the script will randomly select to create realistic-looking login events. The main part of the script is an infinite loop that continuously generates events. For each iteration, it creates a login event dictionary containing a random user ID (between 1-5), the current UTC timestamp, and randomly selected values for IP address, device, and platform. This event is then sent to the "login-events" Kafka topic, followed by a 2-second pause before generating the next event.

Before running the script, always make sure you have "kafka-python" installed. Otherwise, install it by running:

```
pip install kafka-python
```

Then, run the simulator script by

```
python simulator/generate_load.py
```

To consume messages from the topic, run the following command:

```
docker compose exec kafka kafka-console-consumer --bootstrap-server localhost:9092 --topic login-anomalies --from-beginning
```

This command will display all anomalies detected by our Flink job. Initially, you might see many anomalies since all countries will be "new" for each user. After running for a while, anomalies will only be detected when users log in from truly new locations.

> **Hint** To create new anomalies, simply add a new Geo IP to the simulator script and also to the ip_geo.csv file.

The output will look something like this:

```
{
  "user_id": "user_3",
  "timestamp": "2025-06-04T20:15:23.000Z",
  "reason": "Login from new country",
  "country": "DE"
}
```

Each record represents a detected anomaly, showing which user logged in from an unusual location. In a production system, these events could trigger notifications to users or security teams for further investigation.

CHAPTER 8   STREAMING ETL AND ANOMALY DETECTION WITH APACHE FLINK

Summary

Congratulations! You've completed the third and final project focused on streaming data architecture. You now have a functioning anomaly detection system built with Apache Flink.

In this chapter, we built a real-time login anomaly detection system using Apache Flink and Apache Kafka. The system processes streaming login events, enriches them with geographical data, and identifies suspicious login patterns based on unusual locations. We learned how to set up Flink SQL tables, write streaming queries for data enrichment and anomaly detection, and simulate login events for testing.

When deploying the login anomaly detection system to production, several critical considerations must be addressed, with high availability being paramount for both Apache Flink and Kafka components. For Flink, implementing multiple JobManagers with ZooKeeper-based leader election is essential to prevent single points of failure and ensure continuous processing of login events. On the Kafka side, proper cluster sizing with an adequate replication factor (typically 3) across multiple brokers is crucial for fault tolerance and maintaining system reliability. Additional considerations include robust monitoring, security implementations, disaster recovery planning, and thorough documentation of deployment procedures, but these should be addressed only after the core high-availability architecture is established.

This success story demonstrates the tangible impact of real-time streaming architectures in solving real business problems. The reduction in support tickets not only indicates improved security but also translates to cost savings in customer support operations and enhanced customer trust in OneShop's platform. This practical example showcases how stream processing can create immediate business value through proactive problem detection and prevention.

PART III

Machine Learning and Feature Engineering

CHAPTER 9

Building a Product Recommendation Engine with Spark MLlib

The OneShop data engineering team is ready for its next challenge. After successfully building their data lakehouse infrastructure in previous chapters, they're now focused on delivering tangible business value through personalized user experiences. As Priya, the product manager, explains: "We have all this customer data in our lakehouse, but we're not using it to improve how customers actually shop with us."

The immediate challenge involves enhancing product discovery. Currently, the OneShop platform displays the same generic product recommendations to all users, resulting in low engagement and conversion rates. The marketing team has been pushing for a more personalized approach, convinced that tailored recommendations would significantly boost sales.

"What if we could analyze past purchasing patterns and browsing behavior to suggest products customers are actually interested in?" proposes Alex, one of the data scientists. "We already have the data in our lakehouse – we just need to put it to work."

CHAPTER 9 BUILDING A PRODUCT RECOMMENDATION ENGINE WITH SPARK MLLIB

In this chapter, we'll implement a comprehensive product recommendation engine using the Alternating Least Squares (ALS) model from Spark MLlib. This powerful collaborative filtering approach will allow us to generate personalized user and item recommendations based on historical behavior patterns.

Our implementation will utilize the Iceberg lakehouse we built earlier, specifically drawing from the item purchases and pageview tables in the silver layer as our training dataset. The chapter unfolds in three main parts:

First, we'll build a feature engineering pipeline with Spark to transform raw interaction data into meaningful features for our recommendation model, storing these in the gold layer of our lakehouse.

Next, we'll develop a training pipeline to build the ALS model using these features, generating personalized recommendations for all users and persisting them in Postgres for efficient access.

Finally, we'll create a Flask application that serves as an API layer, delivering these recommendations to the frontend application when users browse the OneShop platform.

Before You Begin

You will find the code for this chapter located in the chapter-09 folder. If you haven't set everything up yet, refer to the **Prerequisites** section in the first chapter for more information.

Navigate to the project folder on a terminal by typing:

```
cd <repository_root>/chapter-09
```

CHAPTER 9   BUILDING A PRODUCT RECOMMENDATION ENGINE WITH SPARK MLLIB

Setting Up the Lakehouse Components

Similar to our approach in **Part 1**, we'll continue working with the Iceberg data lakehouse in this chapter. We must ensure our lakehouse infrastructure is operational, as it provides the silver tables we need for feature computation.

The docker-compose.yaml file at the root level is almost identical to the one from **Chapter 3**. The only new addition to the stack is the Flask container, which we'll discuss later in this chapter. If you need a refresher on setting up the lakehouse environment, please refer to **Chapter 3**.

Navigate to the root level of the chapter-09 directory and start all containers by running:

```
docker compose up -d --build
```

This command builds the loadgen and flask containers and launches the entire Docker stack that we explored in **Chapter 3**. This includes:

- Postgres as the source database
- Spark with Iceberg support to run ETL pipelines
- Iceberg REST catalog
- MinIO object storage

Once started, run this script from the root level to start the lakehouse preparation process:

```
./lakehouse-preparer.sh
```

After completion, the lakehouse will be fully populated with data flowing from the bronze to the silver layer.

To verify the contents inside the lakehouse, run the following to launch the Spark SQL shell first:

```
docker-compose exec spark-iceberg /opt/spark/bin/spark-sql
```

CHAPTER 9 BUILDING A PRODUCT RECOMMENDATION ENGINE WITH SPARK MLLIB

Then, run these queries on the shell to ensure the tables have been created and populated.

```
SHOW DATABASES;
SHOW TABLES IN bronze;
SHOW TABLES IN silver;
SHOW TABLES IN gold;

-- Bronze tables

 SELECT COUNT(*) FROM bronze.items;
 SELECT COUNT(*) FROM bronze.pageviews;
 SELECT COUNT(*) FROM bronze.purchases;
 SELECT COUNT(*) FROM bronze.users;

-- Silver tables

SELECT COUNT(*) FROM silver.items;
SELECT COUNT(*) FROM silver.pageviews_by_items;
SELECT COUNT(*) FROM silver.purchases_enriched;
SELECT COUNT(*) FROM silver.users;
```

The script also creates a new Iceberg table in the gold layer with the following schema:

```
CREATE TABLE IF NOT EXISTS gold.als_training_input (
  user_id INT,
  item_id INT,
  interaction_score FLOAT,
  interaction_type STRING,
  feature_ts TIMESTAMP,
  feature_version STRING
);
```

This table is initially empty. It will be populated by running the feature engineering pipeline, which we will discuss next.

CHAPTER 9   BUILDING A PRODUCT RECOMMENDATION ENGINE WITH SPARK MLLIB

Feature Engineering Pipeline

Feature engineering is the process of transforming raw data into features that better represent the underlying problem for the predictive models, resulting in improved model accuracy on unseen data. This critical step in machine learning workflows extracts meaningful information from data sources to create input variables that algorithms can effectively utilize.

For our recommendation engine, we need to build a feature engineering pipeline because the ALS (Alternating Least Squares) algorithm in Spark MLlib requires specific input parameters that are not directly available in our raw data. The algorithm needs numerical user IDs, item IDs, and ratings to learn the latent factors and make recommendations.

We'll transform two silver tables in our Lakehouse – the `silver.purchases_enriched` and `silver.pageviews_by_items` tables – to compute these required features. The pipeline will process this data to extract implicit feedback signals (like purchase history and viewing patterns), convert them to the format expected by the ALS model, and save the computed features in a gold table in the Lakehouse for use in the training process.

Let's examine the key parts of the feature engineering pipeline you can find in the *./spark/scripts/compute_features.py* PySpark script, focusing only on the most important code snippets.

The script starts by creating a Spark session specifically configured for Iceberg tables, which allows smooth interaction with the data lakehouse. It then reads two silver layer tables from the Iceberg catalog: `silver.purchases_enriched` and `silver.pageviews_by_items`. These tables contain purchase history and item viewing patterns that will be processed into features for the ALS model.

CHAPTER 9 BUILDING A PRODUCT RECOMMENDATION ENGINE WITH SPARK MLLIB

The "FEATURE_VERSION" constant suggests that the pipeline includes versioning of features, which is a best practice for model reproducibility and tracking changes in feature engineering over time.

```
from pyspark.sql import SparkSession
from pyspark.sql.functions import col, count, coalesce, lit, current_timestamp
from pyspark.ml.feature import StringIndexer
import logging

FEATURE_VERSION = "v1.0"

# Initialize Spark session with Iceberg support
spark = (
    SparkSession.builder
    .appName("als-feature-extractor")
    .config("spark.sql.catalog.spark_catalog", "org.apache.iceberg.spark.SparkSessionCatalog")
    .getOrCreate()
)

purchases = spark.table("silver.purchases_enriched")
pageviews = spark.table("silver.pageviews_by_items")
```

The pipeline then transforms data through a series of operations to generate meaningful features for the ALS recommendation model. This process converts raw behavioral data into numerical values that represent user-item relationships. These quantified relationships enable the ALS algorithm to learn latent factors and produce accurate recommendations.

```
# ------------------------------------------------
# Aggregate Views
# ------------------------------------------------
view_counts = pageviews.groupBy("user_id", "item_id") \\
    .agg(count("*").alias("view_count"))
```

CHAPTER 9 BUILDING A PRODUCT RECOMMENDATION ENGINE WITH SPARK MLLIB

```
# ----------------------------------------------
# Aggregate Purchases
# ----------------------------------------------
purchase_counts = purchases.groupBy("user_id", "item_id") \\
    .agg(count("*").alias("purchase_count"))

# ----------------------------------------------
# Join Views and Purchases
# ----------------------------------------------
interactions = view_counts.join(purchase_counts,
                                on=["user_id", "item_id"],
                                how="outer") \\
    .withColumn("view_count", coalesce(col("view_count"),
    lit(0))) \\
    .withColumn("purchase_count", coalesce(col("purchase_
    count"), lit(0)))

# ----------------------------------------------
# Compute Interaction Score
# ----------------------------------------------
# You can tune weights: views=1, purchases=3
interactions = interactions.withColumn(
    "interaction_score",
    col("view_count") * lit(1.0) + col("purchase_count") *
    lit(3.0)
)
```

Key highlights are:

1. **Aggregating views**: The pipeline groups pageview data by user_id and item_id, then counts the number of views for each user-item combination.

CHAPTER 9   BUILDING A PRODUCT RECOMMENDATION ENGINE WITH SPARK MLLIB

2. **Aggregating purchases**: Similarly, it groups purchase data by user_id and item_id to count how many times each user purchased each item.

3. **Joining data**: The pipeline performs an outer join between view and purchase counts, ensuring all user-item interactions are captured. For missing values (e.g., items viewed but never purchased), it uses coalesce() to replace nulls with zeros.

4. **Computing interaction score**: Finally, it calculates a weighted interaction score by applying different weights to views (1.0) and purchases (3.0). This weighting reflects the higher importance of purchase actions compared to simply viewing items.

In the final portion of the feature engineering pipeline, two critical steps are performed:

```
# ---------------------------------------------
# Convert IDs to Integer (ALS requires integer IDs)
# ---------------------------------------------
user_indexer = StringIndexer(inputCol="user_id",
outputCol="user_idx", handleInvalid="skip")
item_indexer = StringIndexer(inputCol="item_id",
outputCol="item_idx", handleInvalid="skip")

indexed_model = user_indexer.fit(interactions)
interactions = indexed_model.transform(interactions)

indexed_model = item_indexer.fit(interactions)
interactions = indexed_model.transform(interactions)
```

```
# ---------------------------------------------
# Enrich with metadata for Gold layer
# ---------------------------------------------
final_df = interactions.select(
    col("user_idx").cast("int").alias("user_id"),
    col("item_idx").cast("int").alias("item_id"),
    col("interaction_score").cast("float"),
    lit("composite").alias("interaction_type"),
    current_timestamp().alias("feature_ts"),
    lit(FEATURE_VERSION).alias("feature_version")
)
final_df.write.format("iceberg").mode("overwrite").save("gold.als_training_input")
```

In summary:

1. **Type conversion for ALS compatibility**: The ALS algorithm requires numeric user and item IDs. The code uses **StringIndexer** to convert the string user_id and item_id values into numeric indices (user_idx and item_idx). This transformation is essential as the ALS model can only process integer IDs.

2. **Metadata enrichment**: The final DataFrame is enriched with additional metadata columns. These include:

    - Properly cast integer and float types for the ALS algorithm

    - An interaction_type field set to "composite" to indicate the combined nature of the features

- A timestamp (feature_ts) for when the features were generated
- The feature version for tracking purposes

3. **Persisting to gold layer**: Finally, the transformed data is written to the gold.als_training_input Iceberg table in overwrite mode. This gold layer table will serve as the clean, processed input for the ALS recommendation model training pipeline.

This completes the feature engineering pipeline. Let's go ahead and run it.

```
docker compose exec spark-iceberg /opt/spark/bin/spark-submit \\
  /home/iceberg/pyspark/scripts/compute_features.py
```

Once the script completes, it prepares our data for use in training the recommendation model in the next section.

ALS Model Training Pipeline

After completing the feature engineering process, we'll now move on to training our recommendation model. In this section, we will develop a pipeline that loads the computed features from our gold layer table, trains the ALS model using Spark MLlib, generates personalized item recommendations for all users, and stores these recommendations in PostgreSQL for efficient retrieval by our application.

Let's examine the key parts of the training pipeline you can find in the ./spark/scripts/train_and_serve_als.py PySpark script.

CHAPTER 9   BUILDING A PRODUCT RECOMMENDATION ENGINE WITH SPARK MLLIB

The pipeline begins by loading the features from gold.als_training_input Iceberg table.

```
import sys
from pyspark.sql import SparkSession
from pyspark.ml.recommendation import ALS
from pyspark.sql.functions import current_timestamp, lit, explode, col

POSTGRES_URL = "jdbc:postgresql://postgres:5432/oneshop"
USERNAME = "postgresuser"
PASSWORD = "postgrespw"

MODEL_VERSION = "v1.0"
TOP_N = 5

try:
    spark = SparkSession.builder \\
        .appName("train-and-serve-recommendations") \\
        .getOrCreate()
except Exception as e:
    print(f"Error creating SparkSession: {e}")
    sys.exit(1)

print("SparkSession created successfully.")

# ----------------------------------------------
# Load ALS training data from Iceberg
# ----------------------------------------------
als_input = spark.read.format("iceberg").load("gold.als_training_input") \\
    .filter(f"feature_version = '{MODEL_VERSION}'") \\
    .select("user_id", "item_id", "interaction_score")
```

CHAPTER 9   BUILDING A PRODUCT RECOMMENDATION ENGINE WITH SPARK MLLIB

The code above filters features by MODEL_VERSION to ensure we use only the specific feature version that matches our current model. This filtering is crucial when multiple feature versions exist in the gold layer (from different feature engineering iterations), as it guarantees we only use features that align with our current model architecture and requirements.

Next, the features are passed to the model, which generates the top-N recommendations (five in this case) for all users.

```
# ----------------------------------------------
# Train ALS model
# ----------------------------------------------
als = ALS(
    maxIter=10,
    regParam=0.1,
    userCol="user_id",
    itemCol="item_id",
    ratingCol="interaction_score",
    implicitPrefs=True,
    coldStartStrategy="drop"
)

model = als.fit(als_input)

# ----------------------------------------------
# Generate top-N recommendations per user
# ----------------------------------------------
user_recs = model.recommendForAllUsers(TOP_N)
```

Finally, the pipeline transforms the recommendation output into a structured format and enriches it with metadata.

```
# Flatten recommendations into rows
flattened = user_recs.selectExpr("user_id",
"explode(recommendations) as rec") \\
```

```
    .select(
        col("user_id"),
        col("rec.item_id"),
        col("rec.rating").alias("score")
    )
# Add metadata
enriched_recs = flattened.withColumn("generated_at", current_
timestamp()) \\
    .withColumn("model_version", lit(MODEL_VERSION))
# ----------------------------------------------
# Save to Postgres for real-time serving
# ----------------------------------------------
enriched_recs.write \\
    .format("jdbc") \\
    .option("url", POSTGRES_URL) \\
    .option("dbtable", "user_recommendations") \\
    .option("user", USERNAME) \\
    .option("password", PASSWORD) \\
    .option("driver", "org.postgresql.Driver") \\
    .mode("overwrite") \\
    .save()
```

The pipeline finally writes the enriched recommendations to the user_recommendations table in the source Postgres database. Storing these recommendations in Postgres enables fast retrieval by the application for serving personalized content to users.

We haven't created that table yet. So, let's create it before we run this pipeline.

First, login to Postgres:

```
docker compose exec postgres psql -U postgresuser -d oneshop
```

CHAPTER 9 BUILDING A PRODUCT RECOMMENDATION ENGINE WITH SPARK MLLIB

Let's take a moment to check what's already in this schema.

```
-- See a list of tables
\\dt

-- See sample data from each table

 SELECT * FROM users LIMIT 10;
 SELECT * FROM items LIMIT 10;
 SELECT * FROM purchases LIMIT 10;

 -- Get record counts

SELECT COUNT(*) FROM users;
SELECT COUNT(*) FROM items;
SELECT COUNT(*) FROM purchases;

-- See table structures
\\d users
\\d items
\\d purchases
```

Then, create the table by running:

```
CREATE TABLE user_recommendations (
    user_id INT,
    item_id INT,
    score FLOAT,
    model_version TEXT,
    generated_at TIMESTAMP,
    PRIMARY KEY (user_id, item_id)
);
```

CHAPTER 9   BUILDING A PRODUCT RECOMMENDATION ENGINE WITH SPARK MLLIB

Exit the psql prompt by typing exit

Finally, run the training pipeline.

```
docker compose exec spark-iceberg /opt/spark/bin/spark-submit \\
--jars /home/iceberg/pyspark/jars/postgresql-42.7.6.jar \\
  /home/iceberg/pyspark/scripts/train_and_serve_als.py
```

Note that we included the Postgres JDBC driver jar file location with the script run command. This is necessary because the pipeline needs to write the recommendations directly to Postgres, and the JDBC driver provides the connectivity between Spark and the Postgres database.

You can check the results by running the following query on Postgres.

```
SELECT * FROM user_recommendations LIMIT 10;
```

Serving Recommendations with Flask

Now that we have trained our recommendation model and stored the recommendations in Postgres, we need a way to serve these recommendations to users through our application. To achieve this, we'll implement a lightweight Flask API service that retrieves personalized recommendations for each user from the Postgres database.

Flask is an ideal choice for this task due to its simplicity, flexibility, and low overhead, making it perfect for building microservices that need to handle HTTP requests efficiently. The API will provide a clean interface between our recommendation system and the frontend application, enabling real-time personalized content delivery.

You can find the completed Flask API implementation inside the ./*flask* folder. The app.py file contains the Python implementation code, while the Dockerfile includes instructions to build the custom Flask image and install necessary Python dependencies.

CHAPTER 9   BUILDING A PRODUCT RECOMMENDATION ENGINE WITH SPARK MLLIB

The app.py begins by declaring a function to access the user_recommendations table in Postgres via the psycopg2 library.

```
from flask import Flask, jsonify
import psycopg2

app = Flask(__name__)

# Postgres connection details
POSTGRES_HOST = "postgres"
POSTGRES_PORT = 5432
POSTGRES_DB = "oneshop"
POSTGRES_USER = "postgresuser"
POSTGRES_PASSWORD = "postgrespw"

def get_recommendations(user_id, top_n=10):
    conn = psycopg2.connect(
        host=POSTGRES_HOST,
        port=POSTGRES_PORT,
        dbname=POSTGRES_DB,
        user=POSTGRES_USER,
        password=POSTGRES_PASSWORD
    )
    cur = conn.cursor()
    cur.execute(
        """
        SELECT item_id, score
        FROM user_recommendations
        WHERE user_id = %s
        ORDER BY score DESC
        LIMIT %s
        """,
        (user_id, top_n)
    )
```

CHAPTER 9   BUILDING A PRODUCT RECOMMENDATION ENGINE WITH SPARK MLLIB

```
results = [{"item_id": row[0], "score": row[1]} for row in
cur.fetchall()]
cur.close()
conn.close()
return results
```

Next, the following Flask API code creates a simple endpoint that serves personalized item recommendations to users. The /recommend/ route accepts a user ID as a parameter, retrieves that user's pre-computed recommendations from the Postgres, and returns them as a JSON response. The recommendations are ordered by their score (relevance) and limited to a configurable number (defaulting to 10).

```
@app.route("/recommend/<int:user_id>")
def recommend(user_id):
    recommendations = get_recommendations(user_id)
    return jsonify({"user_id": user_id, "recommendations":
    recommendations})

if __name__ == "__main__":
    app.run(host="0.0.0.0", port=5050)
```

The Flask container is already running as part of our Docker stack. To test our recommendation system, we can use curl to make a request to the Flask API endpoint. Let's see what recommendations we get for user ID 50:

```
curl <http://localhost:5050/recommend/50>
```

Sample response:

```
{
  "recommendations": [
    {
      "item_id": 78,
      "score": 0.00022029372
    },
```

CHAPTER 9   BUILDING A PRODUCT RECOMMENDATION ENGINE WITH SPARK MLLIB

```
    {
      "item_id": 218,
      "score": 0.00013405543
    },
    {
      "item_id": 153,
      "score": 0.000106662614
    },
    {
      "item_id": 451,
      "score": 0.00010095615
    },
    {
      "item_id": 26,
      "score": 0.000081154
    }
  ],
  "user_id": 50
}
```

The response shows the top five recommended items for user 50, ranked by their relevance score. These recommendations were generated by our ALS model based on the user's past interactions and the collaborative filtering algorithm.

Summary

In this chapter, we successfully built a comprehensive recommendation system for OneShop. We implemented a robust feature engineering pipeline that transformed raw interaction data from the Bronze layer into model-ready features in the Gold layer. Using Spark MLlib, we developed and trained an Alternating Least Squares recommendation model

CHAPTER 9   BUILDING A PRODUCT RECOMMENDATION ENGINE WITH SPARK MLLIB

that generates personalized product recommendations based on user interaction patterns. Finally, we created an efficient storage and serving system using PostgreSQL and Flask to deliver these recommendations to users in real-time.

We deliberately separated the feature engineering and model training pipelines for several important reasons. This separation promotes modularity and maintainability, allowing independent updates to either pipeline without affecting the other. It enhances reusability by creating standardized, version-controlled features that could potentially serve multiple downstream models. From an operational perspective, this separation improves efficiency by allowing different scheduling frequencies – features might be updated daily while models could be retrained weekly.

This recommendation system can be extended in several ways. We could implement automation with Apache Airflow to schedule regular pipeline runs using DAGs. An A/B testing framework would allow us to compare multiple model versions simultaneously to continuously improve recommendation quality. We could also enhance the system with real-time feature updates using streaming technologies like Kafka and Spark Streaming. Additionally, adding model monitoring capabilities would help track recommendation quality metrics and model drift over time.

By implementing this recommendation system, OneShop has taken a significant step toward delivering personalized experiences to its customers, leveraging the data lakehouse architecture established in earlier chapters to drive business value through machine learning.

CHAPTER 10

Vector Similarity Search with Postgres and pgvector

Introduction

Throughout this book, we've followed the OneShop's data engineering team on their journey, witnessing their remarkable achievements in building robust data infrastructure. From constructing data lakehouses to implementing ETL processes and designing streaming data architectures, they've established a solid foundation for data-driven decision-making across the organization. Now, the team is ready to embark on yet another exciting project – one that will test their skills in AI application development.

The customer support team at OneShop has approached the data engineering team with a specific need. They want an internal application that allows them to search and analyze customer reviews efficiently. Support managers plan to use this search application to quickly identify reviews containing positive feedback, negative comments, and areas for improvement. This information will be invaluable for understanding

CHAPTER 10   VECTOR SIMILARITY SEARCH WITH POSTGRES AND PGVECTOR

customer sentiment, addressing concerns promptly, and enhancing the overall customer experience. The data engineering team, with its extensive experience in handling various data challenges, is well-positioned to take on this project.

In this chapter, you'll learn how to build a semantic search application using vector embeddings and PostgreSQL with the pgvector extension. We'll walk through the entire process, from setting up the database infrastructure to developing a user-friendly search interface. Along the way, you'll gain insights into:

- The fundamentals of vector embeddings and how they enable semantic understanding of text
- Configuring PostgreSQL with pgvector for efficient similarity searches
- Generating vector embeddings from text using pre-trained models
- Building a streamlined search UI with Streamlit
- Implementing vector similarity search to find conceptually related content

By the end of this chapter, you'll have a functional semantic search application that demonstrates how modern AI techniques can be applied to solve real business problems. This knowledge will equip you to implement similar solutions in your own data engineering projects, especially those requiring natural language understanding capabilities.

Introduction to Vector Embeddings and Similarity Search

In modern AI applications, vector embeddings have emerged as a powerful way to represent data in a format that machines can understand and compare. But what exactly are vector embeddings?

Understanding Vector Embeddings

Vector embeddings are numerical representations of data (text, images, audio, etc.) in a high-dimensional space. Think of them like translating concepts into coordinates on a map – just as Paris and Rome would be positioned closer together on a European map than Paris and Tokyo, in the vector space, similar concepts are positioned closer together. For text data, these embeddings capture semantic meaning by representing words or sentences as dense vectors where related ideas naturally cluster together in this mathematical landscape. For example, in a well-trained embedding space

- The word "king" would be positioned close to "queen," "monarch," and "ruler".
- The sentence "I love this product" would be close to "This item is fantastic" even though they share few words.

The key benefit of embeddings is that they translate complex, unstructured data into structured numerical formats that preserve semantic relationships.

How Similarity Search Works

Once data is converted to vector embeddings, similarity search becomes a mathematical operation to find vectors that are "close" to each other in the embedding space. There are several distance metrics used to measure similarity:

- **Cosine similarity:** Measures the cosine of the angle between vectors (values from −1 to 1, with 1 being identical)
- **Euclidean distance:** Measures the straight-line distance between vectors in the embedding space
- **Dot product:** Another measure of similarity between vectors

For example, if we search for "delicious meal" in a database of restaurant reviews, a similarity search would return reviews with conceptually similar content like "amazing food experience" or "excellent dinner" – even if they don't contain the exact search terms.

Vector Databases: The Infrastructure for AI Applications

As organizations increasingly rely on AI applications that leverage vector embeddings, specialized storage solutions called vector databases have emerged. These databases are optimized for storing, indexing, and querying vector embeddings efficiently.

Traditional databases struggle with vector operations because high-dimensional vectors require specialized indexing techniques, similarity calculations are computationally expensive at scale, and the "nearest neighbor" problem in high dimensions requires specialized algorithms.

CHAPTER 10   VECTOR SIMILARITY SEARCH WITH POSTGRES AND PGVECTOR

Vector databases solve these challenges with specialized indexing structures (like HNSW, IVF, etc.) that make similarity search scalable and efficient.

PostgreSQL and pgvector: Democratizing Vector Search

While several commercial vector database solutions exist (Pinecone, Weaviate, Milvus, etc.), PostgreSQL with the pgvector extension offers an accessible open-source alternative that's gaining significant traction.

pgvector brings vector operations to Postgres by:

- Adding a native vector data type for storing embeddings

- Providing vector similarity operators (<=> for cosine distance, <> for Euclidean distance)

- Implementing efficient indexing methods (IVF and HNSW) for fast similarity search

pgvector's significance lies in its integration of vector search capabilities into a mature, reliable database system that many developers already use, eliminating the need for specialized databases. This makes it an excellent choice for teams building AI applications who want to implement vector search without the added complexity and cost of maintaining separate specialized systems.

Throughout this chapter, we'll see how to leverage PostgreSQL with pgvector to build a semantic search application for customer reviews, demonstrating a practical implementation of these concepts.

CHAPTER 10   VECTOR SIMILARITY SEARCH WITH POSTGRES AND PGVECTOR

Before You Begin

You will find the code for this chapter located in the chapter-10 folder. If you haven't set everything up yet, refer to the **Prerequisites** section in the first chapter for more information.

Navigate to the project folder on a terminal by typing:

```
cd <repository_root>/chapter-10
```

Start the Docker Setup

We don't need the entire data lakehouse infrastructure for this chapter. So we limited the docker-compose.yml file to only two services: postgres and streamlit.

```yaml
services:
  postgres:
    image: pgvector/pgvector:pg16
    container_name: postgres
    ports:
      - "5432:5432"
    environment:
      - POSTGRES_USER=postgres
      - POSTGRES_PASSWORD=postgres
      - POSTGRES_DB=postgres

  streamlit:
    build: ./streamlit
    container_name: streamlit
    init: true
    ports:
      - "8501:8501"
```

```
depends_on:
  - postgres
volumes:
  -./streamlit/embedding_generator.py:/app/embedding_
  generator.py
```

The postgres service uses the pgvector/pgvector:pg16 image, which comes with the pgvector extension pre-installed. The streamlit service builds from our custom Dockerfile located in the *./streamlit* directory, which installs the necessary Python dependencies.

To start these services, run:

```
docker-compose up -d --build
```

Configuring Postgres and pgvector

Let's begin by configuring pgvector on Postgres.

Since we deployed the pgvector/pgvector:pg16 Docker image, pgvector is already pre-installed on Postgres. However, you still need to activate the pgvector extension to make it functional.

Login to Postgres first:

```
docker-compose exec postgres psql -U postgres
```

Then, activate the extension.

```
CREATE EXTENSION IF NOT EXISTS vector;
```

Once activated, you can use "\dx" to see it is listed with others.

```
postgres=# \dx
                              List of installed extensions
  Name   | Version |   Schema   |                Description
---------+---------+------------+-----------------------------------
 plpgsql | 1.0     | pg_catalog | PL/pgSQL procedural language
 vector  | 0.8.0   | public     | vector data type and ivfflat
and hnsw access methods
(2 rows)
```

Insert Sample Reviews

We need some sample data to work with. Let's create the `reviews` table to store customer reviews along with their vector embeddings.

```
CREATE TABLE public.reviews (
    review_id SERIAL PRIMARY KEY,
    customer_name VARCHAR(255),
    customer_email VARCHAR(255),
    date DATE,
    review TEXT,
    review_embedding vector(384)
);
```

The schema of the `reviews` table is a regular Postgres table with standard column types (SERIAL, VARCHAR, DATE, TEXT) except for the `review_embedding` field. This special field is defined as a `vector(384)` data type, which is provided by the pgvector extension.

The 384 value specifies the dimension of the vector. This dimension must match the size of the embedding vectors your model generates. Different embedding models produce vectors of different dimensions – for example, models like MPNet-base produce 768-dimensional vectors, while other models like all-MiniLM-L6 produce 384-dimensional vectors. You must ensure the dimension in your table definition matches the dimension of the vectors your embedding model generates.

Finally, let's insert some reviews into the reviews table. We have a few sample reviews in the "./reviews.sql" file formatted as SQL INSERT statements.

Exit the current psql shell by typing exit. Then, run this command from the root-level of the chapter-10 directory.

```
docker-compose exec -T postgres psql -U postgres
< ./reviews.sql
```

To verify the contents inside the newly populated table, run the following:

```
docker compose exec postgres psql -U postgres -c "SELECT review_id, customer_name, customer_email, date, review FROM public.reviews LIMIT 10;"
```

This will display the first 10 reviews, including the review_id, customer name, email, date, and the actual review text. The query excludes the vector embeddings since they are large numerical arrays that aren't human-readable.

This confirms that our data is properly loaded before we generate embeddings for these reviews.

Search Frontend with Streamlit

Now that our vector database (Postgres with pgvector) is configured and loaded with sample data, the next step is to build the search UI. For this, we will use Streamlit.

Since we already introduced Streamlit in **Chapter 7** when building the flash sales dashboard, we'll skip the detailed explanation here.

CHAPTER 10   VECTOR SIMILARITY SEARCH WITH POSTGRES AND PGVECTOR

You can find the search UI implementation in the `./streamlit` folder. This is where we also keep the *Dockerfile* and the *requirements.txt* files to build the custom Streamlit image and install the following Python dependencies on it.

- **Streamlit:** A Python framework for creating web applications with minimal code
- **psycopg2-binary:** PostgreSQL database driver for Python that allows the search UI to communicate with the Postgres database.
- **sentence-transformers:** Library for generating vector embeddings from text, which we'll use to transform our reviews into embeddings.
- **pandas:** Data manipulation and analysis library

The search UI container is already running in the Docker stack since we ran `docker compose up -d` earlier.

Access the UI by visiting http://localhost:8501/ in your browser. You'll see the following interface:

Figure 10-1 shows the pgvector database schema.

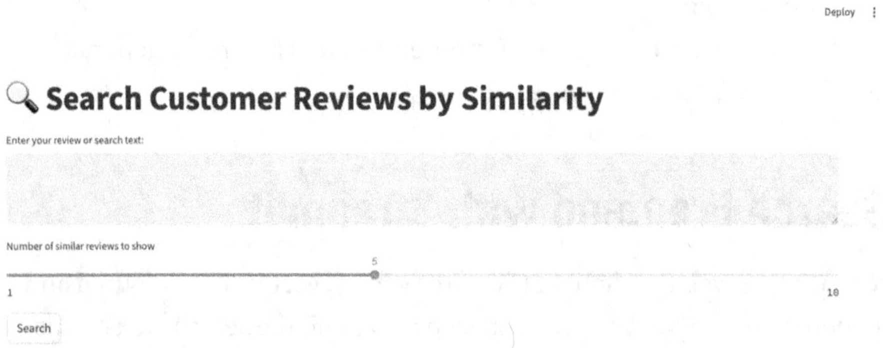

Figure 10-1. Search customer reviews by similarity

The UI is minimal, featuring only a search box and a slider to control how many similar reviews appear in the search results. If you click the "Search" button now, you'll encounter an error.

This error occurs because we haven't yet generated vector embeddings for the reviews. Let's address this next.

Vector Embeddings Generation

The ./streamlit/embedding_generator.py file contains the logic to convert review descriptions into vector embeddings in the reviews table.

```
import psycopg2
from sentence_transformers import SentenceTransformer

# PostgreSQL connection settings
DB_CONFIG = {
    "dbname": "postgres",
    "user": "postgres",
    "password": "postgres",
    "host": "postgres",
    "port": 5432,
}

# Load Sentence Transformer model
model = SentenceTransformer("sentence-transformers/all-
MiniLM-L6-v2")

try:
    conn = psycopg2.connect(**DB_CONFIG)
    cur = conn.cursor()
    cur.execute("SELECT review_id, review FROM public.reviews")
    reviews = cur.fetchall()
```

```
    for review_id, review_text in reviews:
        embedding = model.encode(review_text).tolist()

        # Update the table with generated embedding
        cur.execute(
            "UPDATE public.reviews SET review_embedding = %s
            WHERE review_id = %s;",
            (embedding, review_id),
        )
        conn.commit()
        print("Embeddings updated successfully.")
finally:
    cur.close()
    if conn is not None:
        conn.close()
```

Let's break it down:

The code uses the **Sentence Transformer** library, specifically the all-MiniLM-L6-v2 model. This is a powerful pre-trained model designed to convert text into meaningful vector representations (embeddings). The model transforms sentences into 384-dimensional vectors where semantically similar texts are positioned closer together in the vector space. This particular model is optimized for efficiency while maintaining good performance, making it suitable for applications where computational resources may be limited.

The rest of the code connects to Postgres using psycopg2, retrieves all reviews, generates vector embeddings for each review text using the SentenceTransformer model and finally updates the database with these embeddings. This process transforms the textual content into a format suitable for semantic searching.

Once the embeddings are stored in the database, they can be used with pgvector's similarity search functions to find reviews that are conceptually related to a query, even if they don't share the exact same keywords.

Finally, run the script to generate vector embeddings. This Python script has been mounted to the /app folder of the container.

```
docker-compose exec streamlit python /app/embedding_generator.py
```

Vector Similarity Search

Once the embeddings are generated, what's left here is to walk through the code where the similarity search is performed, which is in the Streamlit application itself.

You can find the fully completed Streamlit application code inside ./streamlit/app.py file.

```python
import streamlit as st
import psycopg2
from sentence_transformers import SentenceTransformer
import pandas as pd

# PostgreSQL connection settings
DB_CONFIG = {
    "dbname": "postgres",
    "user": "postgres",
    "password": "postgres",
    "host": "postgres",
    "port": 5432,
}
```

```python
@st.cache_resource
def load_model():
    return SentenceTransformer("sentence-transformers/all-
    MiniLM-L6-v2")

def find_similar_reviews(query_text, top_n=5):
    model = load_model()
    query_embedding = model.encode(query_text).tolist()

    conn = psycopg2.connect(**DB_CONFIG)
    cursor = conn.cursor()
    cursor.execute(
        """
        SELECT customer_name, review, (review_embedding <=>
        %s::vector) AS similarity
        FROM public.reviews
        ORDER BY similarity ASC
        LIMIT %s;
        """,
        (query_embedding, top_n),
    )
    results = cursor.fetchall()
    cursor.close()
    conn.close()
    return results

st.set_page_config(page_title="Customer Review Search",
layout="wide")
st.title("🔍 Search Customer Reviews by Similarity")

query = st.text_area("Enter your review or search text:", "")
top_n = st.slider("Number of similar reviews to show", min_
value=1, max_value=10, value=5)
```

```
if st.button("Search") and query.strip():
    with st.spinner("Searching for similar reviews..."):
        results = find_similar_reviews(query, top_n)
        if results:
            df = pd.DataFrame(results, columns=["Customer
            Name", "Review", "Similarity"])
            df["Similarity"] = df["Similarity"].apply(lambda
            x: f"{x:.4f}")
            st.dataframe(df)
        else:
            st.info("No similar reviews found.")
```

Let's look at the core functions in this code:

The load_model() function uses Streamlit's caching decorator (@st.cache_resource) to load the sentence transformer model just once and reuse it across multiple searches. This improves performance significantly as loading ML models is computationally expensive.

The find_similar_reviews() function:

- Takes user's query text and desired number of results as parameters

- Encodes the query text into a vector embedding using the same model

- Connects to PostgreSQL database and executes a similarity search using pgvector's cosine distance operator (<=>) to find reviews most similar to the query

- The query sorts results by similarity (lowest distance first) and limits to the top N results

- Returns the matched customer names, reviews, and similarity scores for display in the UI

The rest of the code lays out the search control elements on the search UI. Once the Search button is pressed, the code calls the find_similar_ reviews() function, passing the user's query text and the selected number of results (from the slider). Finally, the result is displayed as a table using Streamlit's dataframe component.

Verify Semantic Search

Now that we have everything set up, let's validate our semantic search functionality. The true power of vector search is its ability to understand the meaning behind words, rather than just matching keywords.

Let's try a sample search using our Streamlit UI:

1. Go to the search UI at http://localhost:8501/

2. In the search box, enter: *"The customer service was very helpful and kind."*

3. Keep the default of five similar reviews or adjust as desired

4. Click the "Search" button

You should notice that the results include reviews about positive customer service experiences, even if they don't use the exact words "helpful" or "kind." For example, you might see reviews mentioning "friendly staff," "great support," or "excellent assistance" – all conceptually similar to your search query.

Figure 10-2 illustrates the similarity search workflow.

CHAPTER 10   VECTOR SIMILARITY SEARCH WITH POSTGRES AND PGVECTOR

Figure 10-2. *Sample search using streamlit UI*

This demonstrates the power of semantic search over traditional keyword search. If we had used a basic SQL LIKE or ILIKE query, we would only find results containing the exact words in our search query. With vector embeddings, the search understands the meaning and returns conceptually similar content.

Try other search queries to explore how the system understands different concepts and returns relevant matches based on meaning rather than exact word matches.

Summary

In this final chapter, we've successfully implemented a vector-based semantic search system for OneShop's customer reviews. We leveraged PostgreSQL with the pgvector extension to store and query vector embeddings and built a user-friendly search interface with Streamlit. This allows the customer support team to find similar reviews based on meaning rather than just keywords, greatly enhancing their ability to analyze customer feedback.

CHAPTER 10 VECTOR SIMILARITY SEARCH WITH POSTGRES AND PGVECTOR

For production deployments, the embedding generation process could be automated in several ways:

- Using Apache Airflow to schedule regular batch processing jobs that generate embeddings for new reviews

- Implementing a real-time embedding generation pipeline using Kafka or similar streaming technologies

- Setting up database triggers to automatically queue new reviews for embedding generation

- Creating CI/CD pipelines to update the embedding model as needed while maintaining backward compatibility

As we reach the end of this book, we'd like to thank you for accompanying OneShop's data engineering team throughout their data platform journey. From establishing foundational data infrastructure to implementing advanced AI applications, you've witnessed the evolution of a modern, data-driven organization. We hope the practical examples and real-world scenarios presented in this book have equipped you with the knowledge and confidence to tackle your own data engineering challenges. Remember that building effective data solutions is an iterative process that requires continuous learning and adaptation.

We wish you success in your data engineering endeavors!

Index

A

Airflow
  admin user creation, 117
  architecture/components, 111
  business impact, 135
  cleanup script, 133
  configuration validation, 117
  connections, 117
  customer segmentation, 109
  DAG (*see* Directed Acyclic Graph (DAG))
  dags directory, 118
  database schema creation, 117
  Docker compose
    configuration, 114–118
  initialization process, 118
  key concepts, 112, 113
  lakehouse infrastructure, 113, 114
  open-source platform, 110
  project root directory, 134
  Trino configuration, 118–124
  TrinoOperator installation, 117
ALS, *see* Alternating Least Squares (ALS)
Alternating Least Squares (ALS)
  feature engineering
    pipeline, 211
  training pipeline, 216
    gold.als_training_input, 217
    Postgres, 219, 221
    recommendations, 216
    structured format, 218
    top-N
      recommendations, 218
Anomaly detection system
  advantages, 189
  containers, 192
  docker-compose.yml
    file, 190–192
  events/verification, 201–203
  login event processing, 193, 194
  low-latency, 188
  services, 192
  SQL tables, 194–201
Apache Software Foundation
  (ASF), 10, 11
Apache Spark
  clean/transform, 77
  comprehensive view, 79
  items table, 80–82
  PySpark (*see* Extract, transform, load (ETL))
  silver tables
    notebook, 84

INDEX

Apache Spark (*cont.*)
    overwritePartitions()
        method, 83
    PyIceberg CLI, 84
    silver.users table, 82
    transformation, 85
    silver.users table, 82
    transformation, 78
ASF, *see* Apache Software
        Foundation (ASF)

B

BI, *see* Business intelligence (BI)
Big data processing, 11
Business intelligence (BI), 99

C

Change Data Capture (CDC),
        140, 165
    anomaly detection system, 188
    configuration, 156
    configuration parameters, 159
    containers, 147
    design pattern, 142
    docker-compose.yml
        file, 143–147
    end-to-end function, 160–162
    feed creation, 148–153
    initialization script, 147
    inventory levels, 147, 148, 161
    key components, 150
    message structure, 153–157

    OpenSearch engine,
        157–160, 162–164
    prerequisites section, 142
    service, 146
    source connector, 149
    transformation configuration, 155
CLI, *see* Command-line
        interface (CLI)
Clickhouse
    analytical queries, 180
    configuration file, 177
    container, 177
    high-performance analytics, 177
    materialized view, 179, 180
    purchases_raw table, 178
    queries, 178
ClickHouse, *see* Real-time analytics
Command-line interface (CLI),
        45, 46, 84
    Trino (Gold tables), 93
    Clickhouse, 177

D

DAG, *see* Directed Acyclic
        Graph (DAG)
Data engineering
    analytical systems, 4, 5
    ASF, 10, 11
    code repository, 18, 19
    definition, 3
    different aspects, 14–16
    Docker environment, 16, 17
    infrastructure, 5

languages, 6
lifecycle
    data format selection, 8
    ingestion, 7
    methodologies, 8
    processing strategies, 8
    serving data, 8–10
    step-by-step process, 6, 7
    storage solutions, 8
OneShop
    activities, 12
    architecture, 13
    data lake serves, 13
    departments, 14
    e-commerce platform, 12
    PostgreSQL Database, 12
    store website, 13
operational systems, 3, 4
performance optimization, 5
quality management, 5
resource requirements, 17
responsibilities, 5
scripting languages, 6
security controls, 5
technologies/languages, 18
Data lakehouse
    adding records, 41, 42
    advantage, 44
    analytics modernization, 22
    architecture, 25, 35
    buckets section, 37
    challenges, 22
    CLI PyIceberg, 45, 46
    components/URLs, 34

    database/iceberg table, 38–41
    definition, 24
    docker-compose.yaml file, 26–29, 34
    docker-compose.yml file, 36, 37
    dual-system approach, 24
    evaluation, 22
    hydrating process, 50
    integration challenges, 21
    Jupyter notebook, 38
    metadata file, 40
    MinIO console login, 36
    modeling iceberg tables, 58–67
    partitioned files, 42
    preparation process, 113
    prerequisites section, 26
    product recommendation engine, 209, 210
    PyIceberg, 42–45
    requirements, 21
    runtime representation, 35
    scalability requirements, 21
    SELECT query, 42
    spark-iceberg Dockerfile, 30–34
    table formats, 25, 26
    warehouses, 23, 24
    vector similarity search, 227
Debezium, 142, 153–157, 174, 175
Directed Acyclic Graph (DAG), 110
    Airflow concepts, 112, 113
    CSV file, 129
    email verification, 133
    engagement categories, 127
    execution, 131, 132

INDEX

Directed Acyclic Graph (DAG) (*cont.*)
 incremental processing, 127
 MinIO bucket, 128
 operation, 128-130
 Python functions, 128
 query filters, 127
 security settings, 124
 source code, 125
 trigger, 130, 131
 TrinoOperator, 125-127
 verification, 132, 133

E

Extract, transform, load (ETL), 4, 24
 containers, 56
 Iceberg (*see* Iceberg tables)
 loadgen service, 55
 MinIO console, 58
 pageviews bucket, 58-60
 postgres and loadgen services, 52-56
 prerequisites section, 52
 PySpark, 50
 transformations, 86
ETL, *see* Extract, transform, load (ETL)

F

Flask
 serving
  recommendation, 221-224

Flink
 anomaly detection (*see* Anomaly detection system)
 concepts, 189
 prerequisites section, 190
 SQL tables
  anomaly detection, 198
  CASE statement, 197
  contents, 196
  data serialization, 195
  declarative language, 194
  enrichment/anomaly detection, 197
  giant squirrel logo, 199, 200
  login anomalies, 196
  login events, 195
  raw login events, 194
  statements, 200
  stream processing, 194
 stateful stream processing, 188, 189

G

Graphical user interface (GUI), 147
GUI, *see* Graphical user interface (GUI)

H

HTTPS, *see* Hypertext transfer protocol secure (HTTPS)
Hypertext transfer protocol secure (HTTPS)

configuration, 118
TLS (*see* Transport Layer Security (TLS))
Trino container, 119

I

Iceberg tables
  Jupyter notebook, 68-71
  lakehouse
    bronze layer, 60-64
    business context, 67
    creation, 58, 59
    data quality, 67
    namespaces, 60
    performance optimization, 67
    silver tables, 64-67
    simplicity/readability, 59
    simplified analytics, 67
  MinIO container, 75
  PageView events, 74-77
  Postgres tables, 72
  PySpark application, 72
  SQL queries, 70
  validation step, 71

J

Java, 19

K, L

Kafka, 157-160
  architecture, 141
  CDC, 155

Clickhouse, 177
concepts, 141
login event processing, 193, 194
OpenSearch, 157-160
streaming data, 141, 142

M, N

Medallion architecture
  Bronze layer, 51
  definition, 51
  Gold layer, 51
  Iceberg tables, 67-76
  layers, 51
  Silver layer, 51
Multi-hop architecture, *see* Medallion architecture

O

OneShop data engineering, 207
  airflow, 109
  benefits, 163
  hydrating process, 49
  key business metrics, 87, 88
  lakehouse (*see* Data lakehouse)
  MinIO data lake, 50
  Postgres database, 50
  real-world data, 12-14
  structured/semi-structured data, 50
  Superset, 87
  vector similarity search, 227
OpenSearch, 157-160
Orchestration solution, *see* Airflow

INDEX

P, Q

PostgreSQL
　Database, 12
　pgvector, 231
　vector (*see* Vector embeddings)
Product recommendation engine
　ALS model (*see* Alternating
　　　Least Squares (ALS))
　feature engineering pipeline,
　　　208, 211–216
　　aggregate views, 212, 214
　　critical steps, 214, 215
　　features, 212
　　gold layer table, 216
　　interaction score, 214
　　metadata columns, 215
　　predictive models, 211
　　silver tables, 211
　　Spark session, 211
　　type conversion, 215
　filtering algorithm, 224
　gold layer, 210
　lakehouse infrastructure, 209, 210
　loadgen and flask
　　　containers, 209
　preparation process, 209
　prerequisites section, 208
　queries, 210
　serving
　　　recommendation, 221–224
　training pipeline, 208
PySpark, 19
Python, 6, 19

R

Real-time analytics, 165
　batch processing, 167
　Clickhouse, 177–181
　components, 167, 168
　connector configuration, 175
　containers, 172
　docker-compose.yml
　　　file, 169–172
　downstream components, 166
　flash sale performance
　　　dashboard, 185
　fundamental building
　　　blocks, 167
　Kafka console consumer, 175
　meaning, 166
　prerequisites section, 168
　purchases table, 174, 175
　Python code, 184
　Redpanda console, 176
　sales dashboard, 166
　source database/schema,
　　　173, 174
　streamlit dashboard, 181–185

S

Semantic search application, 228
Single Message Transforms
　　　(SMTs), 140, 141, 155
SMTs, *see* Single Message
　　　Transforms (SMTs)
Spark MLlib

250

product (*see* Product
 recommendation engine)
SQL, *see* Structured Query
 Language (SQL)
Streaming data, 140, 141, 162
Streamlit
 application code, 239-242
 app.py file, 239
 find_similar_reviews()
  function, 241
 Python dependencies, 236
 sample search, 242
 search customer
  reviews, 236
 search frontend, 235
 sentence transformer
  library, 238
 vector embeddings
  generation, 237-239
Structured Query Language
 (SQL), 6
Superset
 business intelligence, 99
 chart dashboard, 104
 compelling process, 99, 100
 composite dashboard, 106
 configuration, 102, 103
 containers, 91-93
 dashboard, 105, 106
 database connection, 101
 docker-compose.yml file, 88-90
 enhanced dashboard, 107
 operations, 92
 pie chart, 104, 105

prerequisites section, 88
service definitions, 88
SQLAlchemy driver, 100, 101
SQL lab, 103
Trino (*see* Trino
 (Gold tables))

T, U

TLS, *see* Transport Layer
 Security (TLS)
Transport Layer Security (TLS)
 config.properties, 120
 configuration, 119
 Java KeyStore (JKS) file, 121
 password.db file, 120
 security settings, 122
 shared secret, 122
 single-node setup, 122
Trino
 connection, 122-124
 parameters, 123, 124
 TLS/HTTPS
  configuration, 118-124
Trino (Gold tables)
 analytics, 93
 catalogs, 94
 configuration, 95
 conversion rates, 97
 creation, 93, 99
 declarative approach, 93
 iceberg.properties file, 94
 silver tables, 98
 table creation, 96

INDEX

V, W, X, Y, Z

Vector embeddings
  concepts, 229
  configuration, 233
  customer reviews, 227
  databases, 230
  docker-compose.yml file, 232, 233
  indexing structures, 231
  key benefits, 229
  pgvector extension, 228
  postgres/streamlit, 232
  Postgres, 233
  prerequisites section, 232
  insert reviews, 234, 235
  semantic search
    search queries, 243
    verification, 242
  similarity search, 230
  Streamlit, 235–242
Visualization (*see* Superset)

GPSR Compliance

The European Union's (EU) General Product Safety Regulation (GPSR) is a set of rules that requires consumer products to be safe and our obligations to ensure this.

If you have any concerns about our products, you can contact us on

ProductSafety@springernature.com

In case Publisher is established outside the EU, the EU authorized representative is:

Springer Nature Customer Service Center GmbH
Europaplatz 3
69115 Heidelberg, Germany

www.ingramcontent.com/pod-product-compliance
Lightning Source LLC
LaVergne TN
LVHW021956060526
838201LV00048B/1594